VEGETABLE PRIMER

THE SATURDAY EVENING POST

VEGETABLE PRIMER

THE CURTIS PUBLISHING COMPANY
INDIANAPOLIS, INDIANA

The editors wish to thank Marjorie Z. Ashby, Director of Home Economics, and Karen Martin-Quinkert, Home Economist, Stokely-Van Camp, Inc., along with Barb Bacon, Consumer Service Manager, Green Giant Company, for their assistance in compiling this book. A special thanks goes to Charlotte Turgeon, noted editor in the field, for her contributions of unusual recipes and sound advice.

Recipes on the following pages were contributed by Stokely-Van Camp, Inc:
13(r), 73(r), 74(r), 79, 93, 96, 97, 98(r), 107(l).

Recipes on the following pages were contributed by Green Giant Company: 39(r), 44 (l), 46, 104(r), 126, 129(r).

The Saturday Evening Post Vegetable Primer
President, The Curtis Book Division: Jack Merritt
Managing Editor: Jacquelyn S. Sibert
Assistant Editors: Amy L. Clark, Melinda A. Dunlevy
Designer: Pamela G. Starkey
Illustrator: Marcia Mattingly
Proofreader: Kathy Simpson
Technical Director: Greg Vanzo
Compositor: Patricia A. Stricker

Introduction

Note: In this book, an asterisk* before an ingredient indicates it is a recipe within itself and can be found elsewhere in the book. Check the Index under the recipe title for the specific page number.

Contents

What began with select groups of vegetarians, weight-watchers and "health-food nuts" has evolved into a nationwide movement toward better health. Inherent in this movement is a concern for physical fitness and the quality of food intake, which, in turn, has led to dramatic changes in the eating habits of many Americans.

One result of junking the junk food has been a renewed appreciation of foods previously shunned as tasteless, but "good for you." The vegetable is one of these foods. Formerly relegated to the ranks of third-class citizen to the fried chicken and mashed potatoes and cooked to its limpest form, the vegetable is now lauded as a food that is *good*, not just good *for* you.

Raw, cooked "just crisp" or in combination with other ingredients, the vegetable represents a wealth of untapped resources in the realm of taste experiences. With just a little guidance, the vegetable can now run the gamut of the menu, from snacks and hors d'oeuvres to the entrée, or even the dessert course. Its versatile nature allows it to be sometimes light, sometimes elegant, surprisingly substantial—and always nutritious.

This versatility carries over, as well, into the nature of this cookbook. The recipes reflect as much variety as do the vegetables themselves.

There are gourmet dishes like the *Beet Borscht* (23) or *Brussels Sprouts and Chestnuts Veronique* (30), which could either precede an equally elegant entrée or stand as the main course for a special luncheon.

There are classic home-cooked meals like the *All-Day Ham & Bean Soup* (17) and *Baked Stuffed Tomatoes* (117), and vegetables in unusual form, like the braised lettuce on page 81, or the field peas blended into a casserole on page 60.

Cucumber Boats (50) could be the much-talked-about salad at a dinner party, a tasty snack for evening guests or a low-calorie luncheon.

All this, plus an entire section featuring "X-tra Sauces" proven to be exceptional when combined with vegetables, transforms the versatile vegetable into a versatile vegetable cookbook.

The secret to appreciating the main attributes of the vegetable lies in its preparation, which, if done properly, enhances rather than destroys its distinct flavors, colors and textures.

Traditionally, preparation of the vegetable has involved salt. For one, the vegetable was cooked in boiling salted water, which, though it tended to ruin the texture and destroy valuable nutrients, preserved the color and drew out any bitterness in the flavor.

Secondly, salt was often lavished upon vegetable dishes to sharpen the natural flavors. Thirdly, members of the cabbage family, which include the family namesake, as well as broccoli, Brussels sprouts, kale and others, were soaked in salt water to rout out tiny organisms often found inhabiting the leafage.

More recently, however, the use of salt in vegetable preparation has created problems for an increasing number of Americans who are limiting, if not eliminating, salt in their diets. The reasons for this trend are many, varied and controversial, but the fact remains: More and more Americans, by choice or necessity, are reducing their salt intake.

The recipes in this book, therefore, offer substitutes for salt. In cooking the vegetable, the editors suggest the use of lemon juice or a salt substitute, or a selection from the wide range of herbs and spices available. For soaking, we suggest the addition of two tablespoons of white vinegar, rather than salt, to the water.

Of course, some vegetables such as kale, sauerkraut and spinach are naturally high in sodium. And, a few of the recipes in this book call for meats of the pork variety or soy sauce (tamari, in its pure form), which are also high in salt content. Those on a restricted salt diet may wish to bypass these recipes and go on to some of the many others which do not contain these ingredients.

The "sugar question" is handled in much the same way as that of the salt. Little is used in this book, but an occasional recipe will call for brown sugar, or even granulated sugar. In most cases, honey is offered as a substitute. If, however, you are among those who rule out sweeteners in any form, you may wish to select recipes from this book which do not include them.

Make no mistake about it. This is neither a salt-free nor a sugar-free cookbook. It is a book of nutritional recipes which feature the vegetable food class and which offer *alternatives* to the use of salt and/or sugar in vegetable preparation.

It is a book with a built-in flexibility designed to meet the needs of all nutrition-conscious people, while leaving ample room for creative cooking enjoyment.

It is a book which aims to please all—the vegetarian, the "health-food nut," those on restricted diets, those who still yearn for a good ole meat-and-potatoes meal or. . .those who just happen to like vegetables!

The choices are yours to make. So exercise your options and meet the all-new, improved vegetable.

Introduction

A's and B's

The artichoke is one of the oldest cultivated vegetables in the world. It is actually a thistle, of which only the fleshy base of each leaf and the heart are edible. Choose compact, heavy plump globes with large tightly clinging leaf scales of olive green color. Size is not a factor in choosing artichokes.

Artichoke Heart Salad

4 servings

2-3 cups cooked broccoli
1 jar (16-ounce) marinated artichoke
**   hearts**
1 can (11-ounce) mandarin oranges
2 tablespoons grated onion
½ cup coarsely grated carrot
1 tablespoon wine vinegar
3 tablespoons salad oil (olive,
**   peanut or corn)**
Salt and pepper (optional)
Lettuce leaves

Chop the broccoli coarsely in a wooden bowl.

Drain the marinade from the artichoke hearts into a small bowl. Cut the hearts in half and add to the broccoli.

Drain 2 tablespoons of the juice from the oranges into the marinade and add the orange segments to the broccoli.

Sprinkle the broccoli with the onion and carrot.

Add the vinegar and oil to the salad dressing mixture. Season to taste with salt and pepper.

Add the dressing to the salad. Toss well and serve on lettuce leaves.

Cream of Artichoke Soup

4 servings

**4-6 artichokes,
  depending upon size**
⅓ cup hazelnuts
4 cups rich fat-free chicken broth
3 tablespoons rice flour
6 tablespoons cold water
Salt and pepper (optional)
½ cup heavy cream
**1 tablespoon dry sherry
  (optional)**

Remove the top inch and the stem from the artichokes and snip off the prickly tip of the leaves. Cook in boiling water with a teaspoon of lemon juice for about 40 minutes until a leaf pulls off easily. Remove with tongs and drain upside down on paper towels.

Blanch the hazelnuts in boiling water for 10 minutes. Rinse in cold water. Spread on a baking sheet and bake for 10 minutes at 250 degrees F.

Remove the leaves and scoop out the choke from the artichoke. You will only need the hearts. Place the hearts in a pan containing the chicken broth.

Crush the hazelnuts in a blender or food processor or put them in a food grinder. Add to the broth and simmer for 30 minutes. Spin the soup in a blender or food processor and return to the pan.

Mix the rice flour with the cold water. Stir into the soup, gradually bringing it to a boil. Reduce the heat and simmer for 15 minutes.

Season to taste with salt and pepper and stir in the cream and the sherry, if desired. Serve in bouillon cups.

Artichokes

Marinated Artichoke Hearts

6 servings

1 package (9-ounce) frozen artichoke hearts
Vinaigrette*

Place frozen artichoke hearts in a medium saucepan. Pour Vinaigrette* over and around artichoke hearts. Cover and steam for about 15 minutes or until artichoke hearts are heated through, but still firm.

Remove the hearts from the pan; cool and refrigerate for several hours in cooking liquid.

Drain just before serving. Serve on cocktail picks.

Artichokes with Sauterne Sauce

4 to 6 servings

4-6 artichokes

Sauterne Sauce:
1 cup Sauterne
3 tablespoons minced onion
2 cups mayonnaise
3 tablespoons parsley flakes
3 tablespoons lemon juice
1 egg, beaten

Wash and trim artichokes. Cut off about one inch from top and trim tips of leaves. Force the leaves apart and twist out the prickly choke.

Boil in water for 45 mintues or until the outside leaves are easily torn from the artichoke. Drain thoroughly, squeezing out the water by holding the artichoke in a clean dish towel.

Mix together ingredients for sauce and heat slowly. Place a small glass containing the sauce in the center of each artichoke and serve.

Hot Artichoke Dip

4 servings

**1 can (14-ounce) artichoke hearts,
  drained
1 cup mayonnaise
1 cup Parmesan cheese**

Mash the artichokes with a fork and blend
in the mayonnaise and cheese.

Spread in the bottom of a shallow baking
dish and bake at 350 degrees F. for 20
minutes, or until bubbly and golden brown
on top.

Serve hot with crackers of your choice.

Crab Stuffed Artichoke

2 to 3 servings

**1 can (6½-ounce) crab meat, drained
2 ounces Swiss cheese, cubed
1 jar (2-ounce) sliced pimientos, drained
½ cup mayonnaise
2 tablespoons minced onion
¼ teaspoon salt (optional)
2-3 whole fresh artichokes, cooked
  and drained**

Toss the crab, cheese, pimientos,
mayonnaise, onion and salt together.
Remove the small center leaves of the
artichoke, including the hairy center
growth, to leave a cup shape. Fill the
artichoke cups with the crab salad.

Place in 8x8x2-inch baking dish. Pour
hot water in the dish ¼ inch deep around
the artichokes. Cover and bake for 30
minutes at 375 degrees F.

Artichokes

Asparagus is a French and, oftentimes, American delicacy. It is of the lily-of-the-valley family and, like it, comes up in early spring. Select young, tender, green stalks with compact tips. The shorter the time between garden and cooking, the better the product will be.

Asparagus Appetizers

12 to 15 servings

1 pound asparagus
¼ teaspoon sugar
Soy sauce or tamari
Sesame oil

Use the tender tips of the asparagus—the stems will naturally break where the tender part begins. Cut the tips diagonally into 2½-inch pieces.

Drop asparagus into boiling water and cook 2 to 3 minutes. Drain, rinse in cold water and drain again.

Sprinkle sugar, a few drops of soy sauce and a few drops of sesame oil over the asparagus.

Stir to blend seasonings and refrigerate in an airtight container.

Serve cold.

Creamed Asparagus Soup

5 servings

1 cup cooked asparagus
1 cup milk
½ cup cream
1 cup chicken bouillon
3 tablespoons butter
2 tablespoons flour
1 teaspoon salt or ½ teaspoon dried basil
¼ teaspoon white pepper
1 teaspoon Worcestershire sauce
1 tablespoon dry sherry (optional)

Snip tips off asparagus and keep warm. Put stalks and all other ingredients into the blender and spin for 30 seconds or until smooth.

Pour into a saucepan and heat until hot, about 5 minutes.

Add tips to hot soup, stir and serve piping hot.

Sautéed Asparagus

4 servings

1 package (10-ounce) frozen asparagus, thawed
1 tablespoon olive oil
1 tablespoon water
1 tablespoon lemon juice
Carrot curls, for garnish

Cut off asparagus tips. Slice stalks diagonally into thin strips. Heat oil in skillet or wok over high heat. Add asparagus and stir-fry about 30 seconds.

Add the water, cover and steam for 30 seconds more. Uncover, sprinkle with the lemon juice and stir-fry about 30 seconds.

Garnish with carrot curls and serve.

Asparagus

All-American Asparagus

4 servings

1 pound fresh asparagus
Cream Sauce*
⅔ cup grated Cheddar cheese

Cut the hard ends off very fresh asparagus and wash well. Stand the spears in an asparagus cooker with the water coming halfway up the stalks. Steam in salted water for 12 to 15 minutes. To preserve the green color without the use of salt, boil it for one minute, put it in ice water and then steam it or boil it for 10-15 minutes.

Meanwhile make Cream Sauce* and stir in ½ cup of the Cheddar cheese until it melts.

Drain the asparagus well and lay it in a shallow baking dish. Cover it with the sauce and sprinkle the top with the remaining cheese.

Bake at 375 degrees F. until brown (about 5 minutes).

Asparagus Crêpes

4 servings

Cream Sauce*
1 can (8-ounce) water chestnuts, drained
4 asparagus spears, cooked
4 crêpes

Make Cream Sauce* using half-and-half cream in place of milk. Add water chestnuts and asparagus.

Spoon filling into crêpe so that each crêpe gets one whole asparagus spear. Fold into classic roll and serve.

Shelled beans are the developed, but still green, seeds taken from the pod. These include the pinto, garbanzo, red kidney, great northern cranberry, black turtle, Cuban black and large and baby lima beans. Select well-filled pods with beans that are green, but not overly plump.

All-Day Ham & Bean Soup

4 to 6 servings

**1 pound dried Navy beans,
  presoaked and drained**
4 cups water
2-3 pounds ham
1 medium onion, sliced
2 garlic cloves, chopped
1 carrot, sliced
Salt and pepper (optional)
1 package (10-ounce) frozen peas
1 package (10-ounce) frozen limas
½ small head cabbage, shredded

Place all ingredients but frozen vegetables and cabbage in large saucepan or slow cooker. Cover and cook all day (12 to 18 hours in slow cooker).

Remove ham, add peas, limas and cabbage and turn to medium heat (high on slow cooker). Cook for one to 2 hours or until vegetables are tender.

Serve piping hot with a thick crust of bread.

Beans, shelled

Yogurt-Three Bean Salad

8 servings

½ **cup dry garbanzo beans**
½ **cup dry kidney beans**
½ **cup dry black turtle beans**
1 **cup yogurt**
4 **teaspoons lemon juice**
½ **cup nonfat dry milk powder**
1 **tablespoon honey**
¾ **teaspoon garlic powder**
¼ **teaspoon curry powder**
½ **teaspoon basil**
3 **tablespoons fresh chives, chopped**
3 **tablespoons fresh parsley, chopped**

Soak the beans separately overnight. Cook them separately until tender. Drain thoroughly. The canned varieties of these beans may be substituted. They should be well drained. Combine the cooked beans in a nonmetal bowl.

Whisk the yogurt in a small bowl.

Blend the lemon juice with the dry milk to make a thin paste and combine with the yogurt.

Mix the honey, garlic powder, curry powder, basil, chives and parsley and add to the yogurt.

Pour the dressing over the beans. Cover and refrigerate overnight.

Summer Succotash

6 servings

2 **cups cooked lima beans**
4 **cups cooked corn kernels or**
   12 **to 14 ears of corn**
1 **cup milk**
½ **cup cream**
3 **tablespoons butter**
½ **teaspoon sugar**
1 **teaspoon salt or 2 tablespoons**
   minced onion
⅛ **teaspoon white pepper**

Heat the cooked vegetables with the milk and cream, butter, sugar, salt and pepper. Serve in individual dishes.

Boston Baked Beans

4 to 6 servings

20 ounces kidney beans
6 tablespoons brown sugar
½ cup molasses
2 teaspoons dry mustard
2 teaspoons salt or 2 cloves,
  minced
1 medium onion
6 ounces fat salt pork

Soak the beans overnight in a large bowl of cold water. Discard any imperfect ones.

The following morning drain the beans, put them in a kettle, cover with cold water and bring quickly to a boil. Drain the water from the beans, saving the liquid.

Put the beans in a bean pot or in a deep casserole. Stir in the brown sugar, molasses, mustard and seasoning. Poke the onion down into the center. Divide the pork into three pieces. Stick two pieces down into the beans and place the third on top. Add enough liquid to cover the beans, adding more later if necessary.

Bake covered all day, usually 7 to 8 hours, at 275 degrees F. Check occasionally to be sure the beans are still moist, adding more liquid if necessary. Uncover the pot for the last hour of cooking.

Remove the onion before serving the beans.

California Refried Beans with Cheese

4 to 6 servings

20 ounces pinto beans
2 quarts water
1 cup bacon fat
2 cloves garlic, pressed
Salt and pepper (optional)
1½ teaspoons chili powder (optional)
1½ cups diced Cheddar, Monterey Jack
  or coon cheese

Boil the beans in the water over a very gentle heat for 2 hours or until tender. Add more water if needed.

In the bottom of a glazed earthenware pot, an enamel-lined casserole or an old iron pot, heat ⅔ cup of bacon fat. Put in the liquid from the beans and as many beans as it takes to absorb the liquid which you mash with a potato masher. When you have a moist paste, add the rest of the beans, the garlic, salt and pepper to taste and chili powder, if desired. Stir well. The beans are ready to refry at your convenience.

Reheat the beans, adding ⅓ cup of the bacon fat. Stir in the cheese. Stir until well heated and the cheese is melted. The beans should be quite dry, but add a little water if necessary.

Beans, shelled

Snap beans include the green, yellow or wax and Italian beans. The latter is slightly flatter in shape and highly recommended as a new eating experience. With snap beans, pod and all are picked and eaten before the seeds are fully developed. Choose young, tender beans with firm, crisp and slender pods of a good green (or yellow) color.

Hot or Cold Green Beans

4 servings

1 package (10-ounce) frozen green beans
¾ cup olive oil
3 tablespoons wine vinegar
Pepper, to taste
¼ teaspoon basil
½ teaspoon oregano
1 bay leaf
¼ teaspoon garlic powder
¼ cup Parmesan cheese
⅓ cup pignoli nuts

Prepare beans according to directions on package. Drain.

Put oil, vinegar, seasonings and cheese into blender and spin for 10 seconds. Pour into skillet.

Sauté beans in marinade until heated through. Heap beans into serving dish and sprinkle with nuts.

Serve hot or cold.

Snap Bean Salad

6 to 8 servings

2 pounds fresh snap beans
2 small onions, thinly sliced
French Dressing*

Place beans and ½ cup water in large saucepan. Cover and cook 8 minutes over moderate heat. Drain vegetables if necessary and place in a shallow serving dish.

Cover while hot with onion slices and French Dressing*. Spoon the dressing over the beans as they cool.

Refrigerate and serve chilled, but not too cold.

Green Beans with Herbs

6 to 8 servings

2 pounds fresh small green beans
1 teaspoon honey
4 tablespoons butter
1 teaspoon chopped tarragon
2 teaspoons chopped chives
Salt and pepper (optional)

Prepare the beans and put them in a saucepan along with the honey, butter and herbs. (For frozen beans, add one more tablespoon of butter.)

Cover and cook over low heat for 8 minutes. Season with salt and pepper, if desired, and serve.

Beans, snap

The beet is the dark-red root of a leafy vegetable. Its brother, the sugar beet, is white and an important source of sugar. The leaves, or beet tops, are edible as greens. Select beets of uniformly small size, about 1½ inches in diameter.

Mini Beet Kabobs

20 to 30 servings

4 boneless chicken breasts
1 can (16-ounce) whole pickled beets

Place chicken in medium saucepan. Add enough water to just cover the breasts and cook, covered, for 35 minutes over medium heat.

Drain off remaining liquid (keep for use in other recipes) and set chicken out on cutting board to cool.

Drain the beets and chop into bite-size pieces.

When the chicken is cool enough to touch, chop into bite-size pieces.

String a chunk of chicken and a chunk of beet onto toothpicks. Serve at room temperature with pineapple tidbits, sharp Cheddar cheese and plenty of toothpicks.

Cider Beets

4 servings

2 pounds red or yellow beets
⅓ cup light brown sugar
2 teaspoons cornstarch
1 cup hard cider
3 tablespoons butter
Salt and pepper (optional)

Scrub the beets and cut off most of the tops but not the roots. Boil in water for 45 minutes or until tender. Drain and cool in cold water. Slip off the skins and slice into a baking dish.

Preheat the oven to 300 degrees F. Combine the sugar, cornstarch and cider in a small saucepan. Boil for 5 minutes and pour over the beets. Cover and bake 30 minutes. Just before the end of the cooking add the butter in small bits and heat until melted. Season with salt and pepper, if desired, and serve.

Beet Borscht

2 to 3 servings

1 thin slice lemon
1 cup cooked, diced beets
1½ cups sour cream
½ small onion, chopped
½ teaspoon salt or ¼ teaspoon dill
1 cup crushed ice
Sour cream or yogurt

Peel and seed the lemon slice. Place all the ingredients down to but *not* including the ice and sour cream in the blender and spin for 12 seconds.

Add the ice and blend 10 seconds longer, or until smooth and creamy.

Chill for one hour and serve with a dollop of sour cream or yogurt.

Beets

Beet, Cauliflower and Egg Salad

6 servings

1 can (8-ounce) sliced beets
1 head cauliflower
1 small red onion
2 tablespoons chopped parsley
¾ cup Vinaigrette*
3 hard-cooked eggs

Drain the beets and cut into large dice.

Remove the leaves and stem from the cauliflower. Soak it in salted cold water for 10 minutes (see Introduction). Wash, drain and steam 12 to 15 minutes or just until tender. Drain, rinse in cool water and cut off the flowerets with their small stems.

Peel the onion, slice it very thin and divide into rings.

Toss the vegetables with the parsley and Vinaigrette*.

Serve in a shallow salad bowl covered with sliced hard-cooked eggs.

Broccoli is a plant which stands about two feet high. Its edible parts are loosely formed clusters of stunted, under-developed flowers. Try to find freshly picked dark green bunches with no sign of yellow flowers and with tender stalks free from woodiness.

Broccoli Dip for a Party

20 to 30 servings

2 packages frozen chopped broccoli
4 tablespoons butter
1 large onion, chopped
1 can (4-ounce) mushrooms, chopped,
   or ½ cup chopped fresh mushrooms
2 rolls (6-ounce) of garlic cheese
1 can mushroom soup (undiluted)
Salt and pepper (optional)

Cook the broccoli in water until just tender. Drain and set aside.

Melt the butter in a large heavy saucepan. Sauté the onion until translucent and golden but not browned. Add the chopped mushrooms and cook 2 minutes longer.

Add the cooked broccoli, the cheese cut into slices and the mushroom soup to the vegetables in the saucepan. Cook over moderate heat, stirring constantly, just until the cheese melts and blends with the other ingredients. Do not overcook—if you haven't time to watch it closely, cook in the top of a double boiler. Taste and add salt and pepper, if desired.

Put the mixture through a food processor or spin, a little at a time, in a blender. Serve the puree hot, with fresh vegetables or narrow strips of toasted rye bread for dipping.

Broccoli Ring

4 to 6 servings

1 bunch (2 pounds) fresh broccoli or
   2 packages (10-ounce) frozen broccoli
3 eggs, beaten
1 cup mayonnaise
1 cup light cream
1 tablespoon flour
½ teaspoon salt or 2 teaspoons lemon
   juice plus ½ teaspoon oregano

Wash the broccoli, if fresh, and slit the thick stems. Cook in a small quantity of water until just tender. Drain and chop coarsely on a wooden board, discarding any parts of the stems that seem tough or woody. You should have 2½ or 3 cups of the cooked, chopped broccoli.

In a bowl, combine all of the other ingredients. Mix well, then fold in the broccoli.

Pour the mixture into a buttered 1½-quart ring mold. Place the mold in a shallow pan of water and bake 30 minutes at 350 degrees F.

At serving time, turn the mold over onto a serving platter. Run a rubber spatula along the edges to loosen the broccoli mixture, if necessary. Fill the ring with fried shrimp, creamed ham or sliced hard-cooked eggs in cheese sauce.

Broccoli

Chinese Broccoli and Beef

6 servings

2 pounds fresh broccoli
¾ pound beef sirloin
4 tablespoons safflower or corn oil
½ teaspoon minced garlic
2 tablespoons soy sauce or tamari
1 teaspoon sugar
1 tablespoon gin
1 teaspoon salt or ½ teaspoon marjoram
1½ cups water
2 tablespoons cornstarch

Trim the tough stems from the broccoli. Cut the tender stalks and flowerets into small diagonal pieces a little less than 2 inches long. Soak for 10 minutes in salted water (see Introduction). Drain and pat dry with toweling.

Slice the meat, which should be clear of all fat, into thin slices and then into small pieces about the same length as the broccoli.

Heat the oil in a wok or skillet. When very hot, brown the meat, tossing it with two forks, for one minute. Add the garlic, soy sauce, sugar, gin and salt and continue to toss for another minute. Remove the meat with a slotted spoon and set aside.

Add the broccoli and 1¼ cups of water. Cover and bring to a boil. Toss for a moment and cover. Cook 2 minutes.

Add the meat, toss for a moment and pour in the cornstarch mixed with the remaining ¼ cup of water. Cook 2 minutes, stirring gently. Serve on heated plates with hot fluffy rice.

Broccoli Quiche

6 servings

9-inch pastry shell

1 bunch (2 pounds) fresh broccoli or
2 packages (10-ounce) frozen broccoli
4 slices lean bacon
1 medium onion
1½ cups cream
3 eggs
¾ teaspoon salt or 2 teaspoons lemon
juice plus 1 clove garlic, pressed
¼ teaspoon white pepper
Nutmeg
1 cup shredded Gruyère or medium
sharp Cheddar cheese

Prepare your favorite unsweetened pastry shell. Roll it out to a circle 11 inches in diameter and line a deep pie plate. Prick well with a fork and bake 8 minutes at 450 degrees F. Let cool. If this seems too much trouble find an unbaked pie shell in the frozen food section of your local market.

Meanwhile, trim the broccoli of leaves and all but the tenderest of stems and divide into small bunches. Soak the broccoli in salted water for 10 minutes (see Introduction).

Drain and steam or boil in water for 12 minutes or just until tender. If using frozen broccoli cook according to directions on the package, taking care to undercook rather than to overcook. Drain and dry on paper toweling.

Cook the bacon in a small skillet over moderate heat, turning the slices once or twice. Remove the strips to a piece of paper toweling to drain and, using the bacon fat, cook the onion, sliced thin, until tender. Pour off and discard the fat.

Beat the cream and eggs together until well blended and season with salt, pepper and nutmeg. This should be at room temperature when used.

To assemble: Preheat the oven to 350 degrees F. Spread the onion in the bottom of the pastry shell. Arrange the broccoli on top of the onions and sprinkle with the bacon crumbled into small bits. Sprinkle with the cheese and pour the cream mixture over it all. Bake 40 to 45 minutes or until the custard has set and the surface is golden brown.

Broccoli

Brussels sprouts look like tiny cabbages and are, in fact, of the cabbage family. They grow from an erect stalk where one would expect to see leaves. They should be firm and compact with bright, not wilted or yellow, leaves.

Brussels Sprouts Cocktail

1 cup small Brussels sprouts
1 small to medium cabbage
½ pound small cooked shrimp
Scallion Dip*

Drop the sprouts into a pan of boiling water. Bring the water back to a boil and boil 30 seconds. Drain and rinse in cold water. Dry between sheets of paper toweling.

Slice off the root end evenly so that the cabbage will sit gracefully on a serving platter. Peel off the outer layer of leaves. Cut a small bowl-like section out of the top of the cabbage and line with a washed cabbage leaf.

Spear the sprouts and the shrimp with toothpicks and stick them into the cabbage, porcupine fashion, in whatever design you like.

Pour Scallion Dip* into the "bowl." Use the cabbage as a centerpiece for a cocktail party.

Brussels Sprouts Bisque

6 to 8 servings

1 quart Brussels sprouts
2 cups water
1 cup dry white wine
1 medium onion, sliced thin
½ teaspoon salt or ¾ teaspoon
  marjoram
Thyme
1 bay leaf
4 sprigs parsley
4 cups beef broth or canned bouillon
1 cup cream
3 egg yolks
Salt and pepper (optional)
Nutmeg
Toasted slivered almonds

Trim the root ends of the sprouts and remove any wilted leaves. Cut a gash in the bottom of each sprout. Soak in salted water for 10 minutes (see Introduction).

Meanwhile combine the water, wine, onion, salt, thyme, and the bay leaf and parsley, tied in a small bouquet, in a large saucepan. Cover and simmer for 5 minutes.

Drain the sprouts and add to the *court-bouillon* in the pan. Cook uncovered for 10 minutes or until the sprouts are tender.

Remove the bay leaf and parsley from the pan and puree the sprouts and liquid, 2 to 3 cups at a time, until smooth. Strain into the top of a double boiler. Add the beef broth or bouillon and bring to a boil.

Beat the cream and eggs until thoroughly blended and add a little of the hot soup, stirring constantly. Gradually add the mixture to the hot soup and whisk over boiling water until the soup is slightly thickened. Do not let the soup boil. Season to taste with salt, pepper and a dash of nutmeg.

Serve in bouillon cups garnished with the toasted almonds.

Brussels Sprouts

Brussels Sprouts and Chestnuts Veronique

8 servings

1 quart Brussels sprouts
2 cups chicken broth
1 cup dry white wine
1 pound chestnuts
3 tablespoons butter
Salt and pepper (optional)
1 cup seedless white grapes

Remove any wilted leaves from the sprouts and soak them in cold salted water (see Introduction) for 10 minutes. Cut a gash in the bottom of each sprout and boil uncovered in a combination of the chicken broth and white wine for 8 minutes or just until tender.

Drain off the liquid into another saucepan and boil down to ¾ cup.

Meanwhile cut an X on the flat side of each chestnut. Bring a large pan of water to a boil and boil the chestnuts for 20 minutes or until the outer and inner skin can easily be removed with a sharp knife. Taste one to be sure it is edible. If not, boil the peeled chestnuts 5 minutes more.

Chestnuts peel more easily if hot. It is a chore that can be done in advance.

Arrange a mixture of the sprouts and chestnuts in layers in a shallow buttered oven-serving dish, dotting each layer with butter. Sprinkle with the broth and the salt and pepper.

Cover the dish and bake 15 minutes at 350 degrees F. Remove the cover and sprinkle the grapes all over the surface. Bake 8 minutes longer.

Cheesey Brussels Sprouts

4 servings

1 pound Brussels sprouts
3 tablespoons butter
Salt and pepper (optional)
¼ cup grated Cheddar cheese
¼ cup grated Parmesan cheese
¾ cup dry bread crumbs

Cook the sprouts as in the preceding recipe. Drain them well. Toss in the pan with 2 tablespoons of the butter, salt and pepper.

Place them in a shallow oven-serving dish.

Combine the cheese and bread crumbs and sprinkle over the surface. Dot with the remaining tablespoon of butter and brown under a preheated broiler for about 5 minutes.

Brussels Sprouts in Sour Cream

6 servings

1 quart Brussels sprouts or 2 packages
**   (10-ounce) frozen sprouts**
2 cups chicken broth
1½ tablespoons butter
½ pint sour cream
Salt (optional)
¼ teaspoon white pepper
2 teaspoons caraway seeds (optional)

Trim off any wilted leaves and soak the sprouts for 10 minutes in salted cold water (see Introduction). Cut an X in the bottom of each bud. Place the sprouts in a saucepan and add the chicken broth. Partially cover and cook 8 to 10 minutes or just until tender. Drain the liquid into another saucepan and boil it down to ½ cup.

Stir the butter into the second pan.

Add the sour cream, salt, if desired, the pepper and caraway seeds. Mix well off the heat.

Add the sprouts and stir gently until well coated. Put back on the heat but do not boil.

Serve with roast pork, pot roast of beef or roast turkey.

Brussels Sprouts

C's and D's

The cabbage is the head of a large family which includes Brussels sprouts, broccoli, cauliflower, kohlrabi, collards, kale and turnips. All are similar in that the plant has a very short stem and leaves that overlap to form a plump, rounded head. The following recipes deal with the strictly cabbage varieties—red, white, Savoy and so on. The other members are dealt with individually throughout this book. In selecting cabbage, the heads should be reasonably solid and heavy in relation to size, with green outer leaves (except in the case of the red cabbage).

Cabbage and Rice Soup

4 to 6 servings

1 pound green cabbage
2 tablespoons butter
1 large onion, chopped
6 cups chicken broth (canned
  or homemade)
2 cups boiling water
1 cup unpolished long-grain rice
1 teaspoon salt or
  ½ teaspoon dried basil
¼ teaspoon white pepper
1 cup shredded Mozzarella cheese

Wash the cabbage, removing any wilted leaves and the stem. Cut the head into quarters and remove the hard core. Shred coarsely with a long sharp knife or in a food processor. Wash in a large pan of water and dry well in a lettuce drier or between towels.

Heat the butter in a deep ovenproof pan and sauté the onion just until soft. Add 2 cups of the chicken broth and the cabbage. Cover and cook for 10 minutes.

Add the rest of the broth and the water and bring to a full boil.

Add the rice, salt and pepper. Stir well. Cover and cook 20 minutes or until the rice is very tender.

Remove the cover, taste for seasoning and sprinkle with the Mozzarella. Place under a preheated broiler and cook until the cheese melts and is lightly browned.

Bacon Cole Slaw

4 to 8 servings

3 strips bacon
3 tablespoons minced onion
1 cup heavy cream
4 tablespoons cider vinegar
Salt and pepper (optional)
4 cups finely shredded green cabbage

Fry the bacon over moderate heat, turning the bacon twice.

Transfer the bacon to paper toweling and when cool enough to handle crumble quite fine.

Sauté the onion in 2 tablespoons of the bacon fat until soft.

Add the cream and vinegar and stir until blended. Do not boil.

Season to taste with salt and pepper.

Cool and mix with the shredded cabbage. Cover and refrigerate.

Toss in the bacon just before serving.

Cabbage Egg Foo Yong

6 servings

3 cups shredded cabbage
½ cup finely chopped onion
1 teaspoon celery seed
2 cups cooked rice
6 large eggs
¼ teaspoon pepper
2 tablespoons soy sauce or tamari
Safflower or corn oil for frying

Sauce:
2 cups chicken broth
1½ tablespoons cornstarch
1½ tablespoons soy sauce or tamari

Stir together the cabbage, onion, celery seed and rice.

In a large mixing bowl, beat the eggs and then beat in the pepper and soy sauce. Fold in the cabbage-rice mixture.

Prepare the sauce: In a saucepan, combine the chicken broth, cornstarch and soy sauce. Whisk over medium heat until thickened. Keep warm.

To make the patties, pour ⅓-cup portions of the cabbage-rice-egg mixture onto a hot, lightly oiled griddle. Brown on both sides and serve hot, with sauce spooned over each patty.

Cabbage

Cabbage Shrimp Salad Bowl

6 to 8 servings

1 Savoy cabbage
½ pound cottage cheese
3 tablespoons chopped chives
2 hard-cooked eggs, chopped
1 tablespoon tarragon vinegar
3 tablespoons corn or safflower oil
Salt and pepper (optional)
1 pound cooked bay shrimp
Mayonnaise
Parsley
Cherry tomatoes
Black olives

Trim the stem of the cabbage evenly and discard any wilted leaves. Wash well and remove enough of the outer leaves to line a rectangular platter.

Cut the cabbage in half crosswise. Cut out the core and all the center leaves forming bowls with rims approximately ¾ inch thick.

Put the leaves you have removed in a wooden bowl and chop them fine. Stir in the cottage cheese, the chives and hard-cooked eggs. Add the vinegar and oil and mix until well blended. Season to taste with salt and pepper.

Chop half the bay shrimp. Add them to one cup of mayonnaise. Mix well and pile half of the mixture in each cabbage bowl, smoothing it into a dome with a small moistened spatula.

Place the remaining shrimp all over the domes and decorate with parsley flowerets. Using a pastry tube or spoon, run a circle of mayonnaise around the base of each dome. Garnish the ring with the cherry tomatoes and olives.

To serve: Place the bowls on the platter. Cut in wedges. Have a bowl of mayonnaise on the table, for those who want more dressing on their portions.

Polish Pierogi

6 to 8 servings

2 cups flour
½ teaspoon salt or ¼ teaspoon
  celery seed
1 egg
½ cup warm water
1 tablespoon butter
½ onion, chopped
2 cups peeled and cubed potatoes
2 cups shredded cabbage
1 cup cottage cheese
Salt and pepper (optional)
Nutmeg

Heap the flour on a working surface. Fashion a well in the center and put in the salt and egg. Work the egg and salt into the flour with the fingers of one hand while slowly pouring the water with the other until you have a firm ball of dough. Knead the dough on the floured surface until smooth. Let the dough rest in the refrigerator while preparing the filling.

Heat the butter in a small skillet.

Sauté the onion until soft.

Boil the potatoes in water for 10 minutes or until tender. Drain thoroughly and mash.

At the same time, boil the shredded cabbage in ½ cup of water for 10 minutes. Drain well.

Combine the onion, potato, cabbage and cheese. Mix well and season with salt, pepper and a dash of nutmeg.

Divide the dough in half. Roll out one half very thin on a floured surface. Cut into 3-inch circles with a glass or a cookie cutter.

Fill each one with about a tablespoon of the filling. Moisten the edges with your fingertip and fold over like a small turnover, pinching the edges together. Arrange the finished products on a baking sheet. Repeat until all the dough and filling mixture have been used.

Bring a deep skilletful of water to a boil. Poach the pierogi, 10 to 12 at a time, in simmering water for 6 minutes. Remove with a slotted spoon and drain on toweling. Keep warm until all the pierogi are cooked. These may be served immediately with hot melted butter and sour cream or they can be served later, fried in butter and garnished with a small dollop of sour cream.

Cabbage

The carrot is the edible root of a feathery, fernlike plant which has been cultivated as a food for two thousand years. Select firm, well-shaped roots with a good orange color. Buy or pull young carrots, as the large variety are always available in the markets.

Candied Carrots

6 servings

½ **cup water**
2 **tablespoons butter, melted**
1 **tablespoon honey**
1 **tablespoon cornstarch**
1 **tablespoon lemon juice**
2 **cups sliced carrots**
¼ **teaspoon cinnamon**
¼ **teaspoon nutmeg**
⅓ **cup chopped almonds, walnuts**
    or pecans

In medium bowl mix together water, butter, honey, cornstarch and lemon juice.

Stir in carrots, spices and ¼ cup of the nuts.

Cook over low heat in skillet, spooning sauce over carrots frequently, until carrots are tender (about 20 minutes). Be careful not to overcook.

Sprinkle with remaining nuts and serve.

Carrot Soup

5 servings

2 cups grated carrots
1 cup chicken broth
1 cup light cream
1 tablespoon honey
½ small onion, diced
Nutmeg, to taste
Cinnamon, to taste

In a large saucepan combine carrots, broth and cream. Mix well.

Stir in the honey and add the onion and seasonings.

Heat through on medium heat, stirring frequently.

Pour into individual soup bowls and serve with a sprinkle of nutmeg and cinnamon on top.

Country-Time Carrot Bread

1 loaf

1 package (10-ounce) crinkle-cut
**   carrots frozen in butter sauce**
1¾ cups flour
1½ cups sugar
⅓ cup brown sugar
1½ teaspoons baking soda
½ teaspoon cinnamon
½ teaspoon nutmeg
⅓ cup safflower oil
2 eggs, slightly beaten
⅓ cup water
1 teaspoon vanilla

Cook carrots according to the package instructions. Grate them coarsely in a blender, but do not puree. In a large bowl combine flour, sugars, soda, cinnamon and nutmeg.

Stir together carrots, oil, eggs, water and vanilla; add to the dry ingredients and stir until well blended.

Pour the batter into a greased loaf pan and bake for 65 to 75 minutes at 350 degrees F.

Cool for 10 minutes, remove from pan and cool completely before slicing.

Carrots

Carrot Cheese Custard

4 servings

2 tablespoons butter
2 tablespoons chopped onion
2 tablespoons chopped green pepper
2 heaping cups grated carrot
1 cup grated mild cheese
1 tablespoon tarragon vinegar
¾ teaspoon salt (optional)
¼ teaspoon white pepper
1½ cups milk
3 eggs, slightly beaten
Paprika
Parsley

Heat the butter in a small skillet and cook the onion and green pepper over low heat. Combine the onion, pepper, carrots, cheese, vinegar, salt and pepper and mix well.

Preheat the oven to 350 degrees F. Scald the milk and pour slowly into the eggs while beating them with an electric beater or whisk.

Cook the egg mixture over hot water in a double boiler until thick and creamy, stirring constantly. Pour over the carrot mixture and stir until blended.

Bake the custard in a buttered deep casserole dish set in a pan of hot water for 40 to 45 minutes or until a knife inserted comes out clean. Sprinkle with a little paprika and garnish with parsley.

Carrot Juice

4 servings

8 small carrots, cut up
3 cups pineapple juice
1 cup crushed pineapple
2 teaspoons honey
6 ice cubes

Put all ingredients but ice in blender and spin for 5 seconds. Blend in one ice cube at a time.

Garnish with carrot curls and serve immediately.

The cauliflower is a member of the cabbage family, but is grown for its flowers, not its leaves. The stunted underdeveloped flowerets are eaten before the flowers open. A good quality head of cauliflower has white or creamy-white, firm compact curds. Size is not an element in determining quality.

"Cauliflower Cake" Salad

4 to 6 servings

1 head cauliflower
1 cup mayonnaise
2 tablespoons chopped parsley
2 tablespoons chopped chives
3 hard-cooked eggs

Cut off the hard leaves and the center core of the cauliflower. Cut off the flowerets leaving about one inch of the tender stems. Cut a gash in each small stem.

Steam or boil in a little water just until tender. Drain and cool.

Place the cauliflower on a round serving platter. Crush it down slightly by putting a 9-inch layer cake tin on top of it and pressing it gently so that it looks like a thin cake.

"Frost" the cauliflower with ¾ cup of the mayonnaise. Sprinkle generously with the parsley and chives.

Surround the edge with thin egg slices and put a dab of mayonnaise on each egg slice.

Serve cold, cut in wedges.

Cauliflower

Cauliflower Fritters

4 to 6 servings

1 large head cauliflower

Batter:
2 cups flour
1 teaspoon salt or 1 teaspoon lemon
juice plus 1 clove garlic, pressed
1 teaspoon baking powder
1 egg
1 cup flat beer (room temperature)
¾ cup lukewarm water

Cut away all the thick stems of the cauliflower and divide the head into bite-size pieces. Soak for 10 minutes in salted cold water (see Introduction), drain and throw into a pan of boiling water. Bring to a boil and boil 3 minutes. Drain and dry.

Mix the flour, salt, baking powder and egg. Still beating, add the beer and enough water to give the batter a coating consistency rather like all-purpose cream. The amount of liquid will vary slightly with the brand of flour. Cover and let stand.

Just before serving, heat some oil or vegetable shortening in a deep kettle until it registers 380 degrees F. Dip the cauliflower pieces into the batter, which should be at room temperature. Fry 8 or 10 pieces at a time for about one minute or until golden brown. The cauliflower should be tender but firm. Repeat the process until all the cauliflower is fried.

Cauliflower Star

4 servings

1 medium head cauliflower

Sauce:
4 cloves garlic
1 egg
1 teaspoon salt or 1 large clove garlic,
pressed, plus ½ teaspoon marjoram
¼ teaspoon black pepper
1 tablespoon wine vinegar
1 cup olive oil
4 pimiento strips
4 anchovy fillets
12 capers

Cook the cauliflower until just tender.

While it cooks, make the sauce: Mince the garlic and place it in a blender or food processor. Add the egg, salt, pepper and vinegar. Spin 2 seconds and blend in the oil very gradually. The sauce will thicken and as it does the oil can be added a little more rapidly. Taste the sauce for seasoning.

Arrange the cooked cauliflower on a shallow platter. Cover with the sauce. Arrange the pimiento strips and anchovy fillets in a star fashion on the cauliflower and dot with capers.

Celery, of the carrot family, has a watery, crunchy texture and a fresh aromatic odor. There are two varieties of celery available in the United States—the blanched type with white ribs and Pascal, a taller, greener plant. Choose fresh, crisp stalks that are thick and solid with good heart formation.

Braised Celery

2 to 3 servings

3-4 bunches hearts of celery
3 tablespoons butter
3 tablespoons grated onion
3 tablespoons grated carrot
2 cups chicken broth
1 bay leaf
Salt and pepper (optional)
3 tablespoons grated Swiss cheese

Trim the celery evenly and cut lengthwise into quarters. Wash thoroughly. Boil 10 minutes in enough water to cover. Drain well.

Sauté the onions and carrots in the butter and when soft, spread the vegetables in a shallow baking dish. Lay the celery on top of the vegetables. Pour over the chicken broth, add the bay leaf and season with salt and pepper, taking into account the saltiness of the broth.

Cover with aluminum foil and bake 40 minutes at 300 degrees F. Remove the foil and the bay leaf for the last 10 minutes.

Sprinkle with the cheese and serve.

Celery

Layered Celery Salad

8 to 10 servings

1 cup chopped celery
6 cups torn iceberg lettuce
1 can (16-ounce) diagonal-cut green
  beans, drained
4 hard-cooked eggs, sliced
1 cup chopped green pepper
⅓ cup thinly sliced onion rings
1 can (17-ounce) baby peas, drained
2 cups mayonnaise
1 cup grated Cheddar cheese

In a large salad bowl, layer the first seven ingredients in the order given. Spread the mayonnaise over the top of the salad and sprinkle with the grated cheese.

Cover well and let stand for 8 hours or overnight. Serve with a basket of crusty French bread.

Creamed Celery with Nutmeg

2 to 3 servings

4 cups sliced celery
Cream Sauce*
½ teaspoon nutmeg

Boil the celery in 1½ cups of water for 10 minutes. Drain and save the liquid.

Make a well-seasoned Cream Sauce*, substituting one cup of celery broth or one cup of evaporated milk or cream for the milk.

Add the nutmeg and let the celery stand in the sauce for several hours, if possible.

Reheat and serve.

Sweet corn, a member of the grass family, is quite unlike its forerunner Indian corn, or maize. It has even rows of soft, sweet milky yellow or white kernels and is picked before the corn is fully ripe. Select ears with plump, tender kernels and thin, sweet milk. Husks should be green, not dry or yellowish.

Corn Pudding

6 to 8 servings

8 to 10 ears corn
2 teaspoons sugar
2 teaspoons salt or ¼ teaspoon thyme
½ teaspoon basil
¼ teaspoon white pepper
1 tablespoon cornstarch
¼ teaspoon nutmeg
2 cups milk
3 tablespoons butter
3 large eggs, slightly beaten
Parsley

Prepare the corn as for Corn Soufflés*, but to get 2½ cups corn pulp.

Mix the corn with the dry ingredients. Heat the milk with the butter and add gradually to the eggs.

Combine the two mixtures well and pour into a buttered soufflé or casserole dish.

Bake at 350 degrees F. for 50 minutes or until a knife inserted comes out clean.

Sprinkle with chopped parsley and serve at once.

Corn

Cheesey Corn Scallop

6 servings

¼ cup butter
¼ cup flour
1½ cups milk
1 cup shredded Cheddar cheese
1 can (12-ounce) whole kernel corn with
    sweet peppers, drained
1 can (8½-ounce) cream-style corn
3 eggs, beaten
1 tablespoon prepared mustard
1 cup soft bread crumbs
1 teaspoon sugar
½ teaspoon salt or ¼ teaspoon
    marjoram
⅛ teaspoon pepper
½ cup cracker crumbs
1 tablespoon butter, melted

In large saucepan, melt the ¼ cup butter.
Add the flour and stir until smooth.
Gradually add the milk and cheese. Heat
over medium heat, stirring constantly until
sauce thickens.

Add the remaining ingredients, except
for the cracker crumbs and the one
tablespoon butter. Spoon into a 2-quart
greased casserole dish.

Mix the cracker crumbs with the melted
butter and sprinkle on the corn mixture.

Bake uncovered at 350 degrees F. for
one hour or until center is firm.

Magical Quiche

6 to 8 servings

½ pound bacon, crisply fried and
    crumbled
1 cup shredded Swiss cheese
½ cup finely chopped onion
1 can (12-ounce) whole kernel corn with
    sweet peppers, drained
2 cups milk
½ cup biscuit mix
4 eggs
¼ teaspoon salt or ¼ teaspoon
    dried basil
⅛ teaspoon pepper

Combine bacon, cheese, onion and corn;
spread in a greased 10-inch quiche pan or
a deep 10-inch pie pan.

Place remaining ingredients in a
blender. Blend on high speed for one
minute. Pour over mixture in quiche pan.

Bake at 350 degrees F. for 50 to 55
minutes or until a knife inserted comes out
clean.

Allow to set 5 minutes before cutting.

Corn Soufflés

4 to 6 servings

5-6 ears corn
Cream Sauce*
2 tablespoons chopped onion
2 tablespoons chopped pepper
2 tablespoons butter
2 tablespoons chopped pimiento
5 egg yolks, slightly beaten
Salt and pepper (optional)
5 egg whites, beaten stiff

Remove the husks and silks from the ears of corn. Cut off the tips of each ear and scrape off the corn with a sharp knife, to get 1½ cups corn pulp (or use canned corn).

Make the Cream Sauce*, but use 4 tablespoons of butter and 4 tablespoons of flour with the 2 cups of milk in order to make a very thick sauce. Stir in the corn.

Cook the onion and pepper in the butter just until the onion is soft. Add this and the pimiento to the corn and the Cream Sauce*.

Add the beaten egg yolks gradually, stirring constantly, and cook until thick. Season with salt and pepper, if desired.

Remove from the heat. Up to this point the soufflés may be prepared in advance. Thirty minutes before serving preheat the oven to 350 degrees F. Fold the egg whites gently but firmly into the egg yolk mixture.

Spoon the mixture into buttered baking dishes. They should be two-thirds full. Place in the oven and bake 15 to 20 minutes.

(Have an extra buttered baking dish handy in case this is too much mixture for your size dish.)

Corn

Corn and Cheese Bread

6 to 8 servings

3 cups self-rising cornmeal
1 can (8½-ounce) cream-style corn
1 large onion, chopped
3 hot peppers, diced
½ cup grated Cheddar cheese
½ cup safflower oil
½ cup milk
3 eggs, beaten
3 teaspoons sugar
Salt and pepper (optional)

Mix ingredients together well, in the order given.

Pour into a greased shallow baking dish and bake for 30 to 40 minutes at 425 degrees F. or until golden brown on top.

Corn Toss

5 to 6 servings

1 can (16-ounce) whole kernel corn, drained
¼ cup chopped onion
¼ cup chopped green pepper
¼ cup chopped cucumber
¼ cup sweet pickle relish
½ cup Cheddar cheese, cubed
½ cup Thousand Island dressing
5-6 large lettuce leaves
½ cup grated Cheddar cheese, for garnish

In large bowl mix corn, onion, pepper, cucumber, relish and cheese. Fold in the dressing.

Cover and chill for several hours, preferably overnight.

Serve in lettuce cups and top with the grated cheese.

Cucumbers are of the gourd family and, like celery, have a unique flavor, texture and aroma. They are commonly eaten raw, but any recipe calling for cooked squash or gourds can use cooked cucumbers. Select medium-size cucumbers that have a good green color. Avoid those that are too large or that have begun to yellow.

Stuffed Cucumber Luncheon

4 servings

3 boneless chicken breasts
1 large cucumber (about 1½ pounds)
1 teaspoon dill
1 cup cottage cheese
1 teaspoon paprika
5 tablespoons chopped parsley
2 small onions, finely chopped
½ cup chopped walnuts

Place the chicken breasts in a saucepan. Add just enough water to cover them, cover and bring to a boil. Cook for 35 minutes on medium heat. Drain off broth (reserve for use in other dishes, if desired) and place breasts on cutting board to cool.

Meanwhile, wash the cucumber and halve lengthwise. Scoop out the seeds and rub one half with the dill.

Spin the cottage cheese in the blender until smooth, then blend in the paprika and parsley.

Chop the chicken into small pieces and add it and the onions and walnuts to the cottage cheese mixture. Pile this mixture into the unseasoned cucumber half. Place the seasoned half on top and press firmly.

Chill for 20 minutes. Cut into 1-inch slices and serve on a bed of lettuce.

Cucumbers

Cucumber Boats

4 servings

4 cucumbers, 5 inches long
Salt and pepper (optional)
1 pound tiny cooked shrimp
French Dressing*
4 large lettuce leaves
Mayonnaise*
Fresh tarragon leaves

Choose cucumbers of equal size. Peel
them and cut them in two lengthwise.
Scoop out the seeds and discard them.
Scoop out some of the cucumber, leaving
about a ¾-inch shell. Sprinkle the shells
with salt and pepper, if desired.

Dice the scooped-out cucumber and mix
it with the shrimp, which can be bought
cooked and cleaned. Moisten with French
Dressing* made with white wine or
tarragon vinegar.

Pile the shrimp mixture into the
cucumbers. Place each one on a fresh leaf
of garden lettuce. Spoon mayonnaise over
the top and garnish each with 3
tarragon leaves.

Cool Cucumber Soup

4 servings

½ medium cucumber
1⅓ cups plain yogurt
½ teaspoon salt or
    1 teaspoon fresh dill
¼ teaspoon pepper
2 cups buttermilk
2 cups chicken broth
3 hard-cooked eggs, chopped
Paprika

Peel and grate the cucumber into a bowl.
Fold in the yogurt, cover and refrigerate
overnight.

The next day, add the salt and pepper,
buttermilk, broth and 2 chopped eggs. Mix
well.

Serve in individual soup bowls, garnished
with the remaining chopped egg and
paprika.

Cucumber Kabobs

15 to 20 servings

1 medium cucumber
½ pound chunk salami
1 small jar sweet pickle slices, drained
1 small jar dill pickle slices, drained
1 can pitted ripe olives, drained

Wash the cucumber and peel. Cut into strips 3 inches long and ¼ inch wide.

Chop the salami into bite-size chunks.

Place a pickle slice on either side of a salami chunk and put an olive on top. Take skewer and spear a hole into each. Insert cucumber strip "toothpick."

Continue making individual hors d'oeuvres in this manner, alternating pickle flavors (but using 2 of the same on each).

Cucumber Cup Salad

4 servings

2 small cucumbers
Vinaigrette*
4 medium tomatoes
2 small heads garden lettuce
Chopped parsley

Peel and split the cucumbers and scoop out the seeds. Cut in small balls or cubes and cover with a little Vinaigrette*. Place in the refrigerator.

Dip the tomatoes in boiling water for a minute and slip off the skins. Scoop out the centers from the stem ends and turn upside down on a rack to drain.

Wash the lettuce and dry it well. Fill the tomatoes with the cucumbers and place on small individual salad plates lined with lettuce.

Spoon a little more Vinaigrette* over the tomatoes and sprinkle with chopped parsley. Serve very cold.

Cucumbers

E's, F's, G's and H's

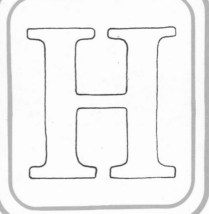

Sometimes known as a brinjal, mad apple or Guinea Squash, the eggplant, of the potato family, is the fruit of a fuzzy green plant with purple flowers. The pear-shaped fruit is usually purple in color and has a waxy, shiny skin. The eggplant should be firm and heavy for its size, with dark purple to purple-black skin.

Fried Eggplant

4 servings

2 medium eggplants
Salt and pepper (optional)
½ cup olive oil
1 large clove garlic, sliced
3 slices onion
2 egg yolks
1 cup fine bread crumbs
Chopped parsley

Cut the eggplant into ½-inch slices. Season with salt and pepper, if desired. Heat the oil in a large skillet and sauté the garlic and onion over moderate heat until both are soft. Remove them from the oil and discard. Pour out half the oil and save.

Meanwhile beat the egg yolks with ¼ cup of water until just blended and place in a shallow dish. Spread the bread crumbs on a plate.

Dip the eggplant slices first in the egg mixture and then in the bread crumbs, coating them well. Prepare as many as the skillet will hold. Fry the eggplant about 4 minutes on each side. Meanwhile prepare the remaining slices. Transfer the first batch to a baking sheet lined with paper toweling. Keep warm and fry the rest, adding more of the fat as needed.

Serve on a hot platter and garnish with fresh chopped parsley.

Eggplant Bake

4 servings

Quick Tomato Sauce*
2-3 eggplants
2 eggs
2 tablespoons water
Fine bread crumbs
Olive oil or salad oil
Salt and pepper (optional)
1 cup grated Monterey Jack or
  or Cheddar cheese

Make the sauce and let it simmer while preparing the eggplant.

Choose eggplants that are not too large. Peel and slice them ½ inch thick, allowing 3 to 4 slices per person. Beat the eggs and water just until blended in a shallow dish. Spread the bread crumbs on a plate. Dip each slice in the egg and then in the crumbs and let them stand on a platter for 10 to 15 minutes.

Preheat the oven to 400 degrees F. Heat enough oil in a large skillet to cover the bottom by ⅛ inch. Brown on both sides as many pieces of eggplant as the skillet can hold. Transfer the browned pieces to a slightly oiled shallow baking dish.

Continue browning and transferring the eggplant, adding more oil to the skillet when necessary. Sprinkle the eggplant with salt and pepper, if desired, and cover with 2 cups Quick Tomato Sauce*. Cover the sauce with grated cheese and bake for 15 minutes.

Eggplant Appetizer

8 servings

4 tablespoons safflower oil
1 medium eggplant
½ cup chopped celery
½ cup chopped onion
1 clove garlic, peeled and minced
1 can (8-ounce) tomato sauce
2 tablespoons white wine vinegar
6 large pimiento-stuffed olives, sliced
Freshly ground pepper, to taste

Heat oil in wok or skillet. Peel eggplant and cut into 1-inch cubes. Stir-fry the eggplant, celery, onion and garlic until tender.

Add the remaining ingredients. Season to taste. Cover and steam for 10 to 15 minutes, stirring frequently.

Let cool and refrigerate. Serve cold as an appetizer.

Eggplant

Real endive has leaves that form a head. The frilled or lacy variety is escarole.

Belgian endive, or witloof chicory, is not a true endive, but a chicory, as the name indicates. The following recipes call for endive, Belgian endive or escarole.

Choose fresh, clean, crisp and cold leaves. Avoid dry, yellowing or wilted leaves or those showing a reddish discoloration of the hearts.

Kale, Spinach and Belgian Endive Salad

8 servings

1 head kale
1 pound fresh spinach
1 head Belgian endive
1 clove garlic
Vinaigrette*
1 cup salad croutons, warmed
Parmesan cheese

Cut off the kale root and most of the white stems. Cut out the center tender leaves and wash them well in cool water. Drain and dry.

Remove the stems from the spinach and wash the leaves in two bowls of water, removing any possible sand. Drain thoroughly in a lettuce drier or between towels.

Cut off the root tip of the endive and separate it into leaves.

Chill all the greens in the refrigerator.

Before serving: Halve the garlic and rub it all around a wooden salad bowl. Add ½ cup of Vinaigrette* and the greens. Toss very thoroughly. Sprinkle with the warm croutons and Parmesan cheese.

Escarole Soup

4 servings

1 small head escarole
1 quart chicken broth
Parmesan cheese

Trim and wash the head of escarole.
Holding the leaves together tightly, cut the
escarole into thin strips. Throw the strips
into a pan of boiling water.

Drain in a colander and run cool water
through the leaves. Stir the escarole into a
pan of hot chicken broth and simmer for 5
minutes.

Serve in soup plates and sprinkle with
freshly grated Parmesan cheese.

Endive and Ham Rolls

4 servings

8 heads Belgian endive
3 tablespoons peanut oil
4 bacon slices
8 slices cooked ham
½ teaspoon salt
¼ teaspoon pepper
¼ cup butter
5 tablespoons whipping cream
Fresh parsley

Wash the endive heads and remove any
damaged outer leaves. Brush four pieces
of foil with the oil.

Cut the bacon into small pieces and
sprinkle over the ham slices. Place one
head of endive on each slice of ham,
season with salt and pepper, roll up and
secure with a toothpick.

Place two ham rolls on each piece of
foil. Dot with butter and marinate each
with the cream.

Seal the foil tightly around each pair of
ham rolls and bake on a baking sheet at
400 degrees F. for 30 minutes.

Remove the rolls from the foil, arrange
them on a platter and serve garnished with
the fresh parsley.

Endive, Escarole

Fiddlehead ferns are found in early spring mostly in northern New England and Canada. They have a flavor similar to asparagus and look like tiny fiddleheads. Markets in Maine sell them occasionally. They are sometimes sold canned or frozen, but mostly are gathered wild. Select young, compact, dark green little sprouts that have not feathered.

Buttered Fiddlehead Ferns

4 servings

3-4 pounds fiddlehead ferns
4 tablespoons butter
Salt and pepper (optional)

Wash the ferns and boil them for 5 minutes in water. Drain them well and toss them in the butter.

Season with salt and pepper, if desired.

Serve warm (or serve them cold in French Dressing*).

Fiddlehead Ferns

Also known as black-eyed peas, cowpeas and black-eyed beans, field peas are the edible seeds of a sprawling plant related to the bean family which is widely cultivated in the South. Pods should be flexible and well filled with tender seeds.

Southern-Style Field Peas

4 to 6 servings

**1 pound dried field peas or
   3 packages frozen field peas
4 cups water
2 teaspoons salt (optional)
¼ teaspoon pepper
1 large onion, chopped
2 stalks celery, chopped
½ pound salt pork, sliced**

Cover the peas with water and soak overnight. Drain and add water. (Add only 2 cups water if using frozen peas.)

Add the remaining ingredients, cover, and cook over low heat for several hours.

At mealtime, serve over mounds of hot, fluffy rice.

Field Peas

Field Pea Casserole

4 to 5 servings

1 pound field peas
1 large onion, chopped
½ cup chopped green pepper
¼ cup diced celery
2 cloves garlic, minced
2 tablespoons salad oil
½ teaspoon black pepper
1 bay leaf
¼ teaspoon thyme
⅛ teaspoon marjoram
2 teaspoons tomato paste
4-5 drops hot pepper sauce
2 pounds lean, cooked ham, cubed
1 cup Burgundy wine

Soak the peas for at least 8 hours in cold water.

Sauté the onion, green pepper, celery and garlic in the oil in a large saucepan or skillet until the vegetables are soft.

Add the drained peas and enough cold water to cover by one inch.

Add the black pepper, bay leaf, thyme, marjoram, tomato paste and hot pepper sauce.

Stir to mix and bring to a boil. Reduce the heat and simmer for one hour.

Add the cubed ham and the wine and continue cooking until the peas are tender and the gravy is thick. Taste for seasoning. If the peas become too dry, add more water.

Serve over cooked rice.

Beet greens, chicory, collard, dandelion greens, kale, mustard greens, spinach, Swiss chard and turnip greens all fall into the category of salad greens. Beet greens and turnip greens are the tops, or foliage, of their more popular root vegetables. Swiss chard is a variety of beet grown for its leaves. The edible parts of collards and kale (the cabbage family) are the ruffled green leaves, while dandelion greens and chicory are the leaves of common weeds. Mustard greens are a family unto themselves. All are treated similarly to spinach, the most popular salad green (other than lettuce). Be sure to choose fresh, young, crisp green leaves. Avoid any with coarse stems or wilted, yellowing leaves.

Chopped Salad Greens in Cheese Sauce

4 to 6 servings

2 pounds greens
3 tablespoons butter
3 tablespoons flour
1¼ cups milk
¾ cup medium sharp Cheddar cheese
½ teaspoon Dijon mustard
Salt and pepper (optional)
4 tablespoons dry bread crumbs

Wash the greens thoroughly, cutting off most of the stems and discarding any bruised or imperfect leaves. Drain and boil for 15 minutes in one cup of water. Drain the greens, pressing out as much liquid as possible. Chop the greens coarsely.

Heat 2 tablespoons of the butter in a small saucepan and stir in the flour. Stir for one or 2 minutes without letting the flour brown. Add all the milk and whisk with a wire whip until the sauce is smooth and thick. Add the cheese and mustard and continue to stir until the cheese is melted. Remove from the heat and season with salt and pepper, if desired.

Stir in the greens and when well mixed put them in a shallow baking dish. Sprinkle with the bread crumbs and dot with the remaining butter. Brown under the broiler and serve.

If this dish is prepared in advance, reheat at 375 degrees F. until the sauce bubbles and the crumbs are golden brown.

Buttered Greens

4 servings

3-4 pounds greens or 2-3 packages
    frozen greens
2 tablespoons vinegar
4 tablespoons butter
⅛ teaspoon nutmeg
Salt and pepper (optional)
2 hard-cooked eggs

Wash the greens thoroughly, picking out any hard stems or wilted leaves.

Place in a large kettle and add one cup of water. Add the vinegar, cover and boil 6 to 8 minutes, depending on the age of the greens. They should be just tender.

Drain in a colander, extracting some of the water with the back of a spoon. Melt the butter in the kettle, return the greens to the kettle and toss just long enough to coat the greens.

Season with the nutmeg and salt and pepper, if desired. Serve in a vegetable dish garnished with wedges of hard-cooked eggs.

Greens

Wilted Dandelion Greens

4 to 6 servings

3-4 quarts dandelion greens
8 slices bacon
1 teaspoon dry mustard
2 tablespoons sugar
6 tablespoons cider vinegar
1 teaspoon salt or ⅛ teaspoon
  nutmeg
¼ teaspoon cracked black pepper

Wash the greens very carefully, discarding the roots and tough stems. Dry between two towels. Tear the greens into small pieces.

Fry the bacon in a large skillet over moderate heat until very crisp. Remove with a bacon fork to paper toweling.

Add the mustard, sugar, vinegar, salt and pepper to the fat in the pan. Bring to a boil and put in the dry greens. Cover and cook the greens for about 3 minutes, until just hot and wilted.

Transfer to a warm vegetable dish and crumble the bacon all over the top.

Fresh Spinach Salad

4 servings

1 package fresh spinach
1 can bean sprouts, drained
8 slices crisp bacon, crumbled
3 hard-cooked eggs, diced
1 cup salad oil
½ cup sugar
⅓ cup ketchup
¼ cup red wine vinegar
1 tablespoon Worcestershire sauce

Tear the spinach into bite-size pieces. In large bowl, toss the spinach, sprouts, bacon and eggs.

Combine the last five ingredients in a small bowl, pour over salad ingredients and toss well.

Spinach Soufflé

4 to 6 servings

3-4 pounds fresh spinach or
   2-3 boxes frozen
3 tablespoons butter
3 tablespoons flour
⅔ cup milk
⅓ cup spinach water
⅛ teaspoon freshly grated nutmeg
4 eggs yolks, slightly beaten
5 egg whites
Hollandaise Sauce*
Paprika

Wash the spinach in several changes of water, discarding the tough stem ends and wilted leaves. Tear into large pieces. Lift the spinach from the final wash water into a kettle. Cover and boil in its own water only 5 to 6 minutes over moderately high heat.

Drain well, reserving ⅓ cup of the spinach water. Squeeze all excess water from the spinach with your fingers. Chop fine.

Heat the butter in a small saucepan and stir in the flour, cooking for 2 minutes without browning. Add the milk and spinach water. Stir until thick and smooth. Season with the nutmeg. Remove from the heat and beat in the egg yolks quickly and thoroughly. Stir in the spinach. This part of the soufflé may be made in advance.

Forty minutes before serving, preheat the oven to 375 degrees F. Beat the egg whites until stiff but not dry. Stir in a third of them until blended. Fold in the remaining egg whites very gently, lifting the mixture high with the whip. Pour into a buttered 1½-quart soufflé dish. Place the dish in a pan of hot water and bake 30 to 35 minutes.

Just before serving cover with Hollandaise Sauce* and sprinkle with a dash of paprika.

Greens

Spinach Openface Sandwiches

4 servings

2-3 pounds fresh spinach
4 slices whole wheat or rye bread
4 slices mild Cheddar or
**   Mozzarella cheese**
Oregano

Remove any tough stems from the spinach, wash the leaves well and place them directly in a saucepan. The water that clings to the leaves will be enough liquid for cooking them. Steam the leaves about 5 minutes, until wilted down and just tender.

Drain off any liquid left in the pan and save it for vegetable stock. Chop the spinach and toast the bread.

Spoon the chopped spinach onto the toasted bread. Top each with a slice of cheese and sprinkle a little oregano on top.

Toast under the broiler until the cheese melts.

Blender Swiss Chard Cream

4 servings

3-4 pounds fresh Swiss chard
3 tablespoons butter
3 tablespoons flour
1 cup milk
½ cup cream
⅛ teaspoon nutmeg
Salt and pepper (optional)
¾ cup buttered croutons

Prepare and cook the Swiss chard as for Buttered Greens*. Save ½ cup of the water.

Place the cooked Swiss chard in the blender. Add the remaining ingredients except the croutons and including the ½ cup cooking water. Spin for 30 seconds.

Place the chard cream in the top of a double boiler or in a baking dish which has been set in a pan of hot water.

Cook 20 to 30 minutes over boiling water (double boiler) or in the oven at 300 degrees F. (baking dish). Garnish with buttered croutons.

Collard Casserole

6 servings

**4 pounds fresh collard greens or
2 packages (10-ounce) frozen
chopped collard greens**
3 hard-cooked eggs
2 tablespoons butter
2 tablespoons minced onion
2 tablespoons flour
1 cup milk
Salt and pepper (optional)
**1 cup medium sharp grated Cheddar
cheese**

Wash the collard greens and strip the leaves from the stems. Boil in a large amount of water until tender. The flavor is strong, so when the leaves are tender (after about 20 minutes), drain and discard the water. Chop the greens coarsely. Or, cook the frozen collard greens according to directions on the box.

Cook the eggs in simmering water for 12 minutes. Cool, shell and slice.

Heat the butter in a small saucepan. Sauté the onions just until soft. Stir in the flour and when blended add the milk. Whisk until smooth and thick. Season to taste with salt and pepper.

Put half the greens in a buttered casserole dish. Spoon half the sauce over the greens and cover with half the sliced eggs and half the cheese. Repeat the process and bake for 20 minutes at 350 degrees F.

For the vitamin-conscious, sprinkle the greens with ¼ teaspoon of powdered Vitamin C (ascorbic acid).

Greens

Hot Turnip Greens Vinaigrette

6 servings

2 pounds turnip greens
3 tablespoons water
Vinaigrette*
1 large clove garlic (optional)

Choose fresh young turnip greens. Wash them well, discarding most of the stems and any bruised leaves. Drain them and place them in a pan, adding the water.

Cover and cook for 10 to 12 minutes or until the leaves are tender.

Sometimes turnip greens can be found in the freezer section of the market. If using the frozen greens, follow directions for cooking, making sure not to overcook.

Drain the cooked greens and while still hot, toss with Vinaigrette*. Add the garlic minced, if desired.

Kale and Beet Salad

4 servings

1 pound kale
1 small can Harvard beets
1 small onion

Dressing:
2 tablespoons red wine vinegar
¼ teaspoon Dijon mustard
½ teaspoon salt or 1 teaspoon
**   grated orange peel**
¼ teaspoon black pepper
3 tablespoons olive oil
3 tablespoons corn or peanut oil
1 hard-cooked egg, chopped

Trim the roots and stems from the kale. Wash the leaves and shred them into a bowl of cold water. Refrigerate.

Slice the beets thin and refrigerate.

Peel and slice the onion very thin. Break into rings and refrigerate in a small bowl of unsalted water.

Put the vinegar, mustard, salt and pepper in a small jar. Cover and shake for a moment or two. Add the oil, shake again and place in the refrigerator.

To assemble: Just before serving, drain the kale and dry it thoroughly in a lettuce drier or between paper towels. Arrange on individual salad plates. Cover with beet slices and then with onion rings.

Shake the dressing and pour over each salad. Garnish with the chopped egg. Serve very cold.

Caribbean Kale Soup

6 servings

1 pound kale
2 tablespoons butter
2 tablespoons minced onion
6 cups chicken broth (canned
   or homemade)
2 cups water
½ pound crab meat
Salt and pepper (optional)

Cut away the tough stems from the kale. Wash the leaves very carefully, removing any bruised leaves.

Heat the butter in a deep heavy pan and cook the onion for one minute.

Shake off any excess water from the kale and put it in the pan. Cover and cook over moderate heat for 10 minutes or until the kale has wilted.

Add the chicken broth and water and simmer uncovered for 20 minutes.

Add the crab meat and simmer 5 minutes longer. Season to taste with salt and pepper, if desired.

Cooked Chicory

4 servings

1 small young head of chicory
4 tablespoons butter
½ cup beef or chicken dish gravy
Salt and pepper (optional)
Croutons

Remove the stem and outer leaves of the chicory. Wash very carefully and boil for 25 minutes.

Drain the chicory and rinse well with cold water. Drain again and chop coarsely.

Heat 2 tablespoons of the butter in a saucepan and toss the chicory in the butter. When the chicory is heated through add the beef or chicken gravy (saved from a roast chicken or roast beef). Heat.

Taste for seasoning.

Serve in a heated vegetable dish and cover with croutons that have been sautéed in the remaining butter.

Greens

I's, J's, K's and L's

The jicama is a Mexican root vegetable of the bean family. It has a very sweet flavor and crisp texture and is used in much the same way as the water chestnut. Choose firm, white bulbs with fresh, green stems.

Chicken and Jicama Crêpes

8 servings

8 crêpes
1 can (10-ounce) cream of chicken or mushroom soup
2 cups cooked, diced chicken
1 cup chopped jicama
¼ cup white wine
Salt and pepper (optional)
Paprika
Worcestershire sauce
Sliced almonds

Prepare the crêpes with a crêpe-maker, set aside and keep warm.

In a medium saucepan, combine the soup, chicken, jicama and wine. Mix well and heat through.

Season to taste with salt and pepper, paprika and Worcestershire sauce. Spoon the mixture onto the center of each crêpe, reserving ½ cup for garnish. Fold into the classic roll.

Top each with a teaspoon of mixture and the sliced almonds.

Jicama

Kohlrabi is a vegetable that has been popular for years in Europe but is just beginning to be widely appreciated in the States. Otherwise known as "cabbage turnip," it is a plant with a bulbous stem growing just above the ground. When young it has edible green leaves. For best flavor the bulbs should be steamed or boiled before they are peeled. Choose young, fresh, firm kohlrabi with crisp, green tops. The older ones tend to be woody.

Creamed Kohlrabi

6 servings

12 kohlrabi
4 tablespoons butter
4 tablespoons unbleached flour
1½ cups whole milk
Salt and pepper (optional)
Nutmeg
½ cup grated Gruyère cheese

Break off the green leaves from the kohlrabi. Cut off any stems and the root. Scrub the bulbs well. Steam or boil until tender.

Wash the leaves and place them in a pan with ½ cup of water. Partially cover and cook until tender. Drain, reserving the liquid. Boil the liquid down to measure one cup. Chop the greens.

Heat 3 tablespoons of the butter in a saucepan and stir in the flour. Cook gently for 2 minutes. Add the milk and vegetable broth. Stir until smooth and thick. Season with salt, pepper and nutmeg. Add the chopped greens.

Peel the kohlrabi and cut into ¼-inch slices. Place in a shallow buttered baking dish. Cover with the sauce. Sprinkle with the grated cheese and dot with the remaining butter. Bake 8 minutes at 400 degrees F.

Kohlrabi

Stuffed Kohlrabi

4 servings

8 medium kohlrabi
¼ pound sausage meat
2 tablespoons minced onion
1 cup mashed potatoes
½ cup soft bread crumbs
Salt and pepper (optional)
2 teaspoons butter
Chopped parsley

Wash the kohlrabi bulbs and peel them with a very sharp knife. Drop them into boiling water and cook whole until tender. Drain and cool.

Cook the sausage meat in a small skillet until light brown. Add the onions and continue cooking until the onions are tender. Drain off all but about one tablespoon of fat. Stir in the potatoes and bread crumbs and salt and pepper to taste.

Trim one end of the kohlrabi bulbs evenly and scoop out the center from the other end. Add the scooped-out portion to the stuffing and mix well.

Place the kohlrabi bulbs in a shallow buttered baking dish. Fill each one with the stuffing and dot with butter.

Bake 20 minutes at 325 degrees F. Sprinkle with chopped parsley before serving.

Kohlrabi Salad

4 servings

8 kohlrabi

Sauce:
1 cup mayonnaise
1 clove garlic, minced
1 tablespoon chopped parsley
1 tablespoon chopped capers
1 hard-cooked egg, finely chopped
2 teaspoons chopped anchovies
½ teaspoon freshly ground black
   pepper

Trim and scrub the kohlrabi bulbs but do not peel them. Boil or steam until just tender. Do not overcook. Drain and plunge into cold water. When cool enough to handle, peel the bulbs, slice them thin and then cut into thin strips. Place in a bowl and chill in the refrigerator.

Combine the ingredients for the sauce. Place in a jar, cover and chill.

Shortly before serving, toss the kohlrabi strips in the sauce. Serve as an appetizer or as a salad on individual plates lined with lettuce leaves.

Leeks belong to the same family as the onion, as do chives, green onions, scallions and shallots. Leeks have long, flat, green leaves and look like slender onions. Green onions are, of course, the tender, underdeveloped sprouts and bulb of the onion. Scallions are really green onions with narrow bulbs, and shallots (not always available in the U.S.) look somewhat like scallions, but grow in clusters of cloves, like garlic. Chives have round, hollow, bright green stems and no bulbs. For all but chives and shallots, select young and tender bunches with fresh green tops and with white coloring extending 2 inches to 3 inches from the bulb base. Chives should be fresh, bright green and not wilted. Choose shallots that are firm with glistening brown skin. (The granddaddy of this group, the onion, is dealt with individually in the next section.)

Quiche Supreme with Leeks

6 servings

4 slices bacon
2 leeks, sliced
1 can (12-ounce) whole kernel corn, drained
1 9-inch unbaked pie shell, well chilled
1 cup grated Swiss cheese
3 eggs, beaten
1 cup half-and-half
½ cup milk
½ teaspoon salt or 1 teaspoon chopped parsley
Pepper, to taste
¼ cup grated Parmesan cheese

Fry the bacon until crisp. Remove to paper toweling and crumble.

Sauté the leeks in the bacon drippings until tender. Remove with a slotted spoon.

Place the corn on the bottom of the pie shell. Top with the cheese and add the crumbled bacon and leeks.

Blend the eggs, half-and-half, milk and seasonings. Pour the egg mixture into the pie shell and sprinkle with Parmesan cheese.

Bake 40 to 45 minutes at 375 degrees F. or until a knife inserted comes out clean. Let stand 5 minutes before cutting.

Leeks

Leeks in White Wine

4 to 6 servings

12 leeks
1 tablespoon diced salt pork
2 tablespoons butter
White pepper, to taste
1½ cups dry white wine
Paprika

Wash the leeks well and trim the stems, leaving two inches of green. Split with knife down to the bulb, but leave attached at the bottom. Wash again. Tie each leek with a little thread to keep them intact while cooking.

Melt the salt pork and butter in a saucepan and sauté the leeks in the drippings until they are browned evenly. Remove what remains of the salt pork and sprinkle the leeks with white pepper.

Add the wine and cover and cook over low heat for 15 minutes. When the leeks are just tender and the wine sauce syrupy, remove the leeks and untie the threads.

To serve, spoon the sauce over the leeks and sprinkle with paprika.

Savory Leeks

4 servings

1 can (16-ounce) cut green beans,
  drained
2 leeks, sliced
3 slices bacon
Salt and pepper (optional)
Slivered almonds

Place green beans, leek slices, bacon and seasonings in medium saucepan. Cover and simmer over medium heat for one hour.

Top with slivered almonds and serve with a slotted spoon.

Leek Soup

4 servings

4 leeks
¼ cup butter
2 medium onions, sliced
2 cups chicken broth
½ teaspoon sugar
1½ teaspoons flour
3 tablespoons water
½ cup grated Gruyère cheese
½ cup cooked ham, cut into strips
1¼ cups cream
Seasoned croutons
Parmesan cheese

Wash the leeks and cut them into 1-inch strips.

Melt the butter in a saucepan and add the leeks and onions. Sauté for 5 minutes. Add the broth and bring to a boil. Simmer for about 5 minutes, then stir in the sugar. Simmer 15 minutes longer.

Mix the flour and water and stir it into the soup. Cook over low heat until the soup thickens, stirring frequently.

Stir in the cheese, ham and cream and heat through, but do not boil.

Serve the soup in individual bowls topped with croutons and a sprinkling of Parmesan cheese.

Scallion Dip

½ cup mayonnaise
½ cup sour cream
1 tablespoon soy sauce or tamari
2 tablespoons ketchup
½ cup finely chopped scallions

In a small bowl, combine mayonnaise and sour cream and stir in the soy sauce.

Mix in the ketchup and chopped scallions. Cover well and refrigerate overnight to allow the flavors to blend.

Serve with Brussels Sprouts Cocktail* or fresh vegetable crudites.

Leeks

Chicken Livers and Green Onions

3 to 4 servings

1 pound chicken livers
½ cup flour
⅛ teaspoon paprika
⅛ teaspoon pepper
5 slices bacon
4 green onions with tops, chopped
4 fresh mushrooms, sliced
1 cup chicken broth
1 can (10-ounce) cream of celery soup
¼ cup dry white wine

Chop the chicken livers into small pieces and toss in flour seasoned with the salt and pepper.

Dice the bacon and fry it to a crisp in a large skillet and drain on paper toweling. Add the chicken livers, green onion and mushrooms to the bacon drippings and sauté until lightly browned.

Pour in the chicken broth, stirring well, and add the bacon bits (reserving some for garnish), celery soup and wine.

Cover and cook over low heat for one hour.

Spoon over mounds of hot fluffy rice and sprinkle with the remaining bacon bits.

Chive Spread

15 to 20 servings

1 cup grated Swiss cheese
1 cup minced cooked ham
1 egg yolk, slightly beaten
1 teaspoon prepared mustard
Salt and pepper (optional)
1 cup chopped chives

Place cheese and ham in medium mixing bowl. Fold in the egg yolk and mustard and season with salt and pepper, if desired.

Form mixture into bite-size balls and roll in the chopped chives.

Arrange on a large platter with whole wheat crackers and serve with warm cider.

Raw Vegetables with Shallot Dip

1 head cauliflower
8-10 tiny new carrots
1 bunch green onions

Dip:
2 shallots
4 ounces smoked salmon
1 package (3-ounce) cream cheese
½ cup heavy cream
¼ cup mayonnaise
⅛ teaspoon white pepper
1 tablespoon chopped parsley

Remove the leaves and coarse stem from the cauliflower. Using a very sharp small knife, cut the cauliflower into bite-size flowerets, leaving a thin stem on each to serve as a handle. Soak the flowerets in salted cold water (see Introduction) for 30 minutes. Drain and chill in the refrigerator.

Scrub the tiny carrots but do not peel them. Cut off all but 2 inches from the green onions and trim the root end. Chill both the carrots and the green onions.

Put the shallots and smoked salmon in a blender or food processor and spin until minced. Add the remaining ingredients and process until smooth. Place the mixture in a low decorative bowl.

To serve, place the bowl containing the dip in the center of a serving platter, preferably not a white one. Surround the dip with circles of cauliflower and accent at random with the carrots and green onions. Serve *cold*.

Leeks

Lentils are delectable legumes that look like miniature peas, except they are flat. The seed of lentils split like peas. Split Egyptian lentils, as they are called, are easier to find than whole lentils, which are sold as brown lentils. The latter ones are, however, more flavorful.

Lentil Soup

4 to 6 servings

2 cups lentils, pre-soaked and drained
8 cups water
1 medium onion, chopped
2 stalks celery, chopped
1 carrot, sliced
5 slices bacon
1 clove garlic, minced
2½ teaspoons salt or 2 teaspoons
  fresh chopped basil
¼ teaspoon pepper
½ teaspoon oregano
1 can (16-ounce) whole tomatoes
Parmesan cheese

Combine in a large saucepan lentils, water, onion, celery and carrot.

Dice the bacon and fry it to a crisp in a skillet. Drain on paper toweling and crumble. Reserve ¼ cup bacon bits for garnish. Add the rest, plus the garlic, seasonings and tomatoes to the vegetables.

Mix well, cover and simmer for 4 to 6 hours.

Serve in individual soup bowls, sprinkled with the bacon bits and Parmesan cheese.

Lentil Ragout

8 servings

½ cup safflower oil
⅓ cup thinly sliced onion
½ cup chopped green pepper
1 pound boneless pork shoulder, cut in
   1-inch cubes
1 can (16-ounce) sliced carrots, drained
1 can (16-ounce) tomatoes
1 can (15-ounce) tomato sauce
1½ cups lentils, rinsed
1 teaspoon salt or 2 teaspoons
   fresh chopped basil
⅛ teaspoon freshly ground black pepper
1 bay leaf
Ground cloves

Heat the oil in a large saucepan.

Add the onion and green pepper and cook until tender, stirring frequently. Remove the vegetables with a slotted spoon and set aside.

Brown the pork in the remaining drippings. Add more oil if necessary to keep the meat from getting too dry. When the pork is lightly browned, drain off the excess fat.

In a large saucepan, combine the cooked vegetables, carrots, tomatoes, tomato sauce, lentils, salt, pepper, bay leaf and cloves. Add the meat.

Bring to a boil. Reduce the heat, cover and simmer for about 1½ hours or until the lentils are just tender. Stir the mixture occasionally as it cooks.

Additional water may be added if the mixture seems dry.

Lentils

Lettuce, of the chicory family, comes in three types. In head lettuce, the leaves grow in a round, hard-packed cluster like cabbage; in leaf lettuce, the leaves grow in a loose bunch, and in romaine lettuce, the leaves grow like leaf lettuce but are long and narrow. Head lettuce should be firm, but "give" slightly when squeezed. Leaf and romaine lettuce should be fresh, crisp and bright green. Avoid leaves that are wilted, yellowing or that have brown spots.

Lettuce Salad Rolls

4 servings

1 head fresh garden lettuce
2 cups mixed cooked vegetables
2 tablespoons minced onion
French Dressing*
1 cup cottage cheese
Mayonnaise
Radishes

Wash the lettuce well, separating the leaves. Lay the leaves out on a towel in pairs, one leaf on top of another.

Mix the vegetables and onion with ½ cup of dressing, the cottage cheese and just enough mayonnaise to bind the vegetables together.

Spread the mixture on the lettuce leaves, leaving a margin of one inch all around. Fold the sides of the lettuce toward the center and then roll up the lettuce leaves. Place seam side down on a serving platter, cover with pliofilm or foil and refrigerate for at least one hour.

Before serving pour a little dressing over each roll and garnish with sliced radishes or radish roses.

Lettuce Basket

6 servings

1 head fresh garden lettuce
2 avocados
1-2 cucumbers
Cherry tomatoes
Black olives

Dressing:
⅓ cup olive oil
⅓ cup safflower oil
⅓ cup red wine vinegar
1 large clove garlic, minced
2 tablespoons chopped chives or
    green onion tops
½ teaspoon salt or lemon juice
½ teaspoon black pepper

Wash the lettuce thoroughly and drain
well. Pat dry the leaves with toweling and
pile the leaves one on top of the other. Roll
the leaves up and cut into thin strips with a
large knife. Fluff the strips with your hands
and line a salad bowl with the lettuce.

In another bowl combine the ingredients
for the dressing. Peel and seed the
avocados and cut in slices into the
dressing.

Peel and halve the cucumber(s) and
scoop out the seeds with a spoon. Slice
very thin into the dressing.

Just before serving transfer the
vegetables to the lettuce bowl. Garnish
with halved cherry tomatoes and pitted
olives and toss in the dressing.

Braised Lettuce

6 to 8 servings

2 large bunches leaf lettuce
4 tablespoons butter
Sugar
2 cups chicken broth
1 tablespoon vinegar
Salt and pepper (optional)

Wash the lettuce well and place 3 to 4
large leaves one on top of the other. Roll
them up and tie them securely with kitchen
twine. Trim them evenly with kitchen
scissors or, if too large, cut them in two.
Continue to do this until you have 2 rolls
per serving.

Drop the rolls into boiling water and boil
3 minutes. Retrieve with a slotted spoon
and drain on toweling.

Place the rolls in a shallow baking dish
greased with one tablespoon of the butter.
Sprinkle with a little sugar. Add the chicken
broth, 3 tablespoons butter and vinegar
and sprinkle with salt and pepper, if
desired.

Cook covered for 30 minutes at 350
degrees F. Remove the cover and cook for
10 minutes more.

Serve immediately.

Lettuce

M's, N's, O's and P's

The versatile mushroom is not a true vegetable, but a fungi. It has no leaves, blossoms or fruit and comes in many shapes and colors. It is very low in calories. The freshest mushrooms are closed around the stem by a thin veil and may be white, tan or cream-colored. Those having open veils due to loss of water are just as nutritious and have a more pungent flavor.

Mushroom Crêpes

6 to 8 servings

6-8 crêpes
½ pound fresh mushrooms
4 tablespoons butter
3 green onions, chopped
3 tablespoons flour
1 cup half-and-half
¾ cup sour cream

Prepare the crêpes with a crêpe-maker and set aside.

Separate the mushroom stems from the caps and brush off the sand. Chop enough stems to equal ½ cup. Set aside. Slice the remaining stems and the caps.

Melt the butter and sauté the sliced mushrooms and onions. Stir in the flour and gradually add the half-and-half. Cook over moderate heat, stirring constantly until thick and smooth.

Add ½ cup of the sour cream. Spoon the mixture onto the center of the crêpes and fold into the classic roll. Top each with a dollop of the remaining sour cream and sprinkle with the diced mushrooms.

Mushroom Appetizers

25 to 30 servings

1 pound fresh mushrooms
2 cups biscuit mix
½ cup cold water
¼ pound ground sausage
¼ cup finely chopped chives or
    green onions
1 package (3-ounce) cream cheese,
    softened
2 cups grated Cheddar cheese
Paprika

Remove the stems from the mushrooms and brush the sand off both the stems and the caps. Finely chop the stems, but leave the caps whole. Set aside.

Combine the biscuit mix with the water to form a soft dough. Mix it well (about 20 strokes). Flour hands to keep the dough from sticking and press the dough into the bottom of a greased 13x9x2-inch baking pan.

Brown the sausage in a skillet, separating it with a wooden spoon as it cooks. Drain off the excess fat. Mix the chives, cream cheese and chopped stems into the sausage.

Fill the mushroom caps with the sausage mixture. Arrange the stuffed mushrooms in rows on the dough in the pan and sprinkle the cheese and paprika all over the top.

Cover loosely with foil and bake for 20 minutes at 350 degrees F. Remove the foil and bake 5 to 10 minutes longer, or until the cheese is brown and bubbly.

Let stand for 15 minutes before cutting.

Mushrooms

Stuffed Mushrooms

18 to 24 servings

18-24 large mushrooms
¼ pound Canadian bacon
4 tablespoons butter
2 tablespoons chopped onion
4 tablespoons sherry
½ teaspoon dried basil
Salt and pepper (optional)
Fine bread crumbs
Parmesan cheese

Cut off the ends of the stems and remove the caps, wiping them clean. Lay the mushroom caps in a shallow, buttered baking dish. Chop the stems and cut the bacon into small pieces.

Heat the butter in a small skillet and cook the onion, chopped mushrooms and bacon all together until the mixture is quite dry and lightly browned. Add the sherry, basil and salt and pepper, if desired, and continue cooking until the sherry has evaporated.

Stuff the mushroom caps with the mixture. Cover each cap with the bread crumbs and with Parmesan cheese. Dot each one with butter and bake 20 minutes at 400 degrees F. Serve as a garnish around a roast or on toast for a luncheon dish.

Mushrooms Sous le Verre

6 to 8 servings

1½ pounds mushrooms
6-8 slices firm white bread
6 tablespoons butter
1 tablespoon chopped parsley
1 teaspoon chopped fresh tarragon or
⅓ teaspoon powdered dried tarragon
2 teaspoons lemon juice
¼ teaspoon black pepper
1½ cups cream
3 tablespoons Madeira or sherry

Remove the stems from the caps. Wipe the caps clean. Cut the slices ½ to ¾ inch thick from a loaf of white unsliced bread, preferably homemade. Toast them.

Butter the center of six to eight salad-size ovenproof serving plates. Stir the herbs, lemon juice and pepper into the butter. Spread on each piece of toast. There will be a little left over.

Place a toast on each plate. Put the mushroom caps on the toasts and cover with 2 tablespoons of cream. Place the plates on two baking sheets and cover them with glass bells or Pyrex bowls. Bake 30 minutes at 350 degrees F.

Combine the rest of the cream with the wine and add any herb butter that remains. Heat but do not boil. Just before serving remove the bell and spread a spoonful of sauce over the top. Replace the bell and serve.

The long pointed green pods which follow the red and yellow blossoms of the okra plant are what we eat as okra and what the southerners use in gumbo. Buy pods that are young, green and tender and no longer than 3 inches. Larger pods are stringy and the seeds are hard.

Creole Okra

4 to 6 servings

1¼ pounds small okra
3 tablespoons bacon fat
1 medium onion, chopped
1 medium green pepper, chopped
3 cups tomatoes, peeled and diced
1 clove garlic, pressed
1 tablespoon brown sugar
1 teaspoon lemon juice
1 bay leaf
1 teaspoon salt (optional)
¼ teaspoon freshly ground black pepper

Wash the green pods well, removing their stems. Slice them. You should have about 2½ to 3 cups.

Heat the fat in a heavy skillet or pot and add the prepared vegetables and garlic, removing the seeds from both the pepper and tomatoes. Cover and simmer 5 minutes.

Add the brown sugar, lemon juice, bay leaf, salt, pepper and okra and stew 30 minutes or until the okra is tender. Serve very hot.

Okra

Fried Okra

4 servings

1 pound fresh okra
½ cup cornmeal
⅛ teaspoon pepper
¾ teaspoon dill
½ cup safflower oil

Wash the pods well and cut off the ends.
Slice into little wheels. Dip each wheel into
the cornmeal and sprinkle with pepper and
dill.

    In a skillet, heat just enough oil to cover
the bottom. Fry the coated wheels until
brown and crunchy on both sides, adding
more oil as needed.

    Drain on paper toweling and serve.

The onion is a bulb plant of the lily family.
The bulb is protected by a thin, paper-like
skin that is peeled off before eating.
Spanish onions are brown skinned and of
medium sharpness. Bermuda onions are
white and sweet and red onions are red
and sweet. Most onions are sold dried or
"cured." Choose onions that are clean and
firm. The skins should be dry, smooth and
crackly. Avoid onions with wet, soggy
necks or soft, spongy bulbs.

Sweet Onions

4 to 6 servings

14-16 small onions
4 tablespoons butter
2½ tablespoons brown sugar
Salt and pepper (optional)

Peel the onions and pierce each one with the sharp tines of a small fork. Cover with water and boil gently for 15 minutes, or until tender. Drain and dry on toweling for 30 minutes, turning them occasionally.

Heat the butter in a skillet. Add the sugar and cook one minute. Toss the onions in the mixture until well coated on all sides. Season to taste. Place in a shallow baking dish and bake uncovered for 20 to 30 minutes. Serve as a garnish or in the baking dish.

Creamed Onions

4 to 6 servings

1½-2 pounds medium onions
Cream Sauce*
½ cup heavy cream
¼ teaspoon nutmeg
½ cup fine bread crumbs
Butter
Chopped parsley

Peel the onions, holding them under cold water as you peel to keep from getting teary. Boil them in water for 15 minutes or just until tender. Do not overcook. Drain thoroughly.

Make the Cream Sauce*. Add the cream and season with nutmeg. Both the onions and sauce can be made a day in advance, but combine them just before the final baking.

Place them in a buttered baking dish. Cover with bread crumbs and dot with butter. Bake 30 minutes at 325 degrees F. Garnish with chopped parsley.

Onions

Quick Onion Rolls

24 rolls

½ **cup butter**
2 **tablespoons minced onion**
1 **tablespoon instant beef bouillon**
1 **teaspoon parsley flakes**
¼ **teaspoon onion powder**
2 **cups biscuit mix**
½ **cup cold water**

In a medium saucepan, mix together the butter, onion, bouillon, parsley flakes and onion powder. Warm over medium heat until the butter is melted and the bouillon is dissolved. Cool slightly.

Pour about half of the butter mixture into a round layer pan and spread the onion evenly in the pan.

Blend the biscuit mix with the water until it forms a soft dough. Mix well (about 20 strokes).

Drop the dough by teaspoonfuls onto the butter mixture in the pan. Dribble the remaining butter mixture over top.

Bake at 425 degrees F. for about 12 minutes or until golden brown. Invert the pan onto a serving platter and let sit for a few minutes before removing and serving.

Parsnips look somewhat like carrots and are of that family, but they are white and more slender. The leaves of the parsnips are coarse, not feathery and fernlike as in the carrot. Choose smooth, well-shaped parsnips that are tender and free from woodiness. They should be of small to medium size. Do not use the very large variety. Discoloration in "fresh" parsnips may be an indication of freezing.

Buttered Parsnips

4 to 6 servings

2 pounds parsnips
4 tablespoons butter
⅛ teaspoon fresh nutmeg
Salt and pepper (optional)
1 tablespoon chopped chives

Wash, trim and scrape the parsnips. Cut into uniform pieces and boil in water 25 to 30 minutes or until tender. Drain well and let dry.

Just before serving, heat the butter in a skillet and sauté the parsnips over moderate heat until light brown on all sides, letting them caramelize a little in their own sugar.

Season with nutmeg and salt and pepper, if desired. Serve in a warm vegetable dish garnished with chopped chives.

Creamed Parsnips

4 to 6 servings

2 pounds parsnips
1 small onion, sliced
Cream Sauce*
1 tablespoon chopped parsley

Scrub or scrape the parsnips according to their age and size. Cut into thick slices or leave whole if they are very young. Boil covered in 2 cups of water along with the onion for 15 to 20 minutes or until tender. Drain. Remove the onion and place the parsnips in a buttered casserole dish.

Make the Cream Sauce*, taste for seasoning and pour over the parsnips.

Bake for 15 minutes at 325 degrees F. Garnish with fresh parsley and serve.

Parsnips

The pods of garden peas, in the pea family, appear after the small white flowers have faded from the twining stems and leaves of the plant. The snow pea, a variety of this plant, is dealt with in the next section. Select fresh green garden-picked pea pods which are well and uniformly filled. The pods should look crisp and unwrinkled. The peas within the pods should be green and firm.

Fresh Pea Puree

4 servings

**2 pounds fresh peas or 1 package
   (10-ounce) frozen peas
2 medium potatoes
2 tablespoons butter
1 package (3-ounce) cream cheese
¼ cup sour cream
Salt and pepper (optional)
Buttered croutons
Bacon bits**

Shell the peas if fresh. Boil fresh or frozen peas in ½ cup of water in a covered saucepan for 10 minutes. Remove the cover and let the peas cook almost dry without allowing them to scorch. Place the peas in a food processor or blender.

Meanwhile, peel and quarter the potatoes and steam or boil them until tender. Drain and add the potatoes to the peas. Add the butter, cream cheese and sour cream. Spin until smooth. Taste for seasoning and add salt and pepper, if desired.

Divide the puree between four individual soufflé dishes or ramekins. Heat the ramekins in a pan of boiling water for 25 minutes at 325 degrees F. Serve piping hot, topped with croutons and bacon bits.

Peas à la Blue Cheese

4 servings

1 can (17-ounce) sweet peas
2 tablespoons butter
¼ cup crumbled blue cheese
Caraway seed

Drain the peas, reserving ½ cup of the liquid. In a small saucepan, heat the peas and the liquid until hot.

In another pan, melt the butter and stir in the blue cheese.

Drain the liquid off the peas, add the blue cheese mixture and stir lightly to blend.

Sprinkle with a dash of caraway seed and serve.

Saucy Peas

5 servings

2 tablespoons butter, melted
2 tablespoons flour
1 can (17-ounce) sweet peas
¼ cup milk
¼ teaspoon salt or ¼ teaspoon thyme
**   plus 2 teaspoons chopped parsley**
4 ounces pasteurized process cheese,
**   cubed**
1 can (8-ounce) water chestnuts,
**   drained and sliced**

In a medium saucepan, combine the butter and flour. Drain the peas, reserving the liquid. Stir the pea liquid and milk into the saucepan, gradually. Cook over medium heat, stirring constantly until thick and bubbly.

Add the salt and cheese, stirring until the cheese melts.

Fold the water chestnuts and peas into the sauce mixture and heat to serving temperature.

Peas

Peppers are of the nightshade family and are red, yellow or green in color. The sweet or bell pepper is quite mild. The Anaheim or California pepper varies from sweet to mildly hot. Jalapeño peppers are hot and serrano peppers are *very* hot. Pimientos are a special type of sweet red pepper grown in southern climates. They have a subtle sweet, yet distinct, flavor and should be used with discretion. Select peppers which are fresh, firm and thick-fleshed with bright green, red or yellow coloring. Immature peppers are usually soft and dull looking.

Hot Pepper Relish

1 pint

2 sweet green peppers
2 sweet red peppers
1 stalk celery, chopped
1 medium onion, chopped
¾ teaspoon salt or salt substitute
¼ cup sugar
¼ cup vinegar
1 teaspoon red pepper flakes
⅛ teaspoon cinnamon

Seed the peppers and chop coarsely. In medium saucepan, combine all ingredients and mix well. Bring to a boil and simmer for 30 minutes.

Cool and serve with party bread and cold cuts.

Peppers and Chicken Livers

6 servings

1 pound chicken livers
4 tablespoons flour
1 teaspoon salt (optional)
½ teaspoon pepper
1 teaspoon paprika
1 cube (1-inch) fresh ginger
1 clove garlic
3 tablespoons corn or safflower oil
3 medium green peppers
2 teaspoons cornstarch
½ cup plus 1 tablespoon water
2 tablespoons sherry
2 tablespoons soy sauce or tamari
3 cups cooked rice

Chop the chicken livers into bite-size pieces. In a shallow bowl, combine the flour, salt, pepper and paprika. Coat the liver pieces in the mixture, then rub the mixture into the meat with your fingers. Shake off the excess and set aside.

Peel the ginger and peel and crush the garlic. Heat 2 tablespoons of the oil in a skillet or wok. Add the ginger and garlic and stir-fry until the garlic is browned. Remove and discard both.

Seed the green peppers and cut into ½-inch strips. Add the pepper to the skillet or wok and stir-fry about one minute. Cover and steam for 2 minutes, stirring frequently, then stir-fry again until tender, but crisp. Remove and set aside.

Add the remaining one tablespoon of oil to the skillet or wok and stir-fry the liver pieces, a few at a time, until golden brown. Add more oil, if needed. Remove the liver strips and set aside.

Dissolve the cornstarch in the one tablespoon of water. Add the remaining water, sherry and soy sauce to the skillet or wok and stir in the dissolved cornstarch.

Add the pepper strips and liver back in and stir-fry until the sauce thickens and the peppers and liver are heated through.

Serve over generous mounds of hot, fluffy rice.

Peppers

Buffet Party Pimiento Loaf

6 to 8 servings

½ cup French Dressing*
2 cans (10½-ounce) chicken broth
1 package (3-ounce) lemon-flavored
   gelatin
1 envelope unflavored gelatin
Onion juice
Red pepper sauce
Salt and pepper (optional)
2 jars (4-ounce) whole pimientos,
   drained
1 package (8-ounce) cream cheese,
   softened
2 cups chopped celery
½ pound thinly sliced chicken
½ pound thinly sliced ham

Heat the French Dressing* and broth together. Add both gelatins, stirring until dissolved. Add the seasonings. Chill until thickened and syrupy.

Cut desired designs from pimientos with a cookie cutter or sharp knife to decorate the top of the loaf. Stir the cream cheese, celery and bits of pimiento left from designs together.

Coat the inside surface of a large loaf pan with enough thickened gelatin to cover the bottom. Arrange a third of the pimiento cutouts on the gelatin and chill until set.

Fill the mold with the remaining ingredients in separate layers, starting with a third of the chicken slices, a third of the ham slices, a third of the pimiento designs and a third of the cheese mixture. Spoon a third of the gelatin mixture over all, letting the gelatin fill into any spaces to level off a layer. (Cut the chicken and ham slices very thin and pierce with a fork so the gelatin can seep through for easier slicing of the mold.)

Repeat the layers two times, ending with the gelatin. Chill overnight.

At party time, slice and serve with an assortment of breads.

Pimiento Chicken Rolls

4 servings

4 boneless chicken breasts, skinned
½ cup flour
4 thin slices baked ham
1 jar (4-ounce) whole pimientos, drained
4 slices Swiss cheese
4 tablespoons butter
¾ cup dry white wine
Salt and pepper (optional)

Sauce:
1 tablespoon minced onion
¼ cup dry white wine
1 cup sour cream
2 egg yolks
Salt and pepper (optional)

Rub the outside of the chicken with the flour. Place a slice of ham, half of a pimiento and a slice of cheese on one side of each breast. Fold the second side over to cover the filling and secure with toothpicks.

Sauté the chicken rolls in the butter until golden brown on all sides, about 20 minutes. Drain off the excess fat.

Add the wine and sprinkle the breasts with salt and pepper, if desired. Cover and cook over medium low heat for 45 minutes, or until tender. Remove from the pan.

To prepare the sauce, simmer the onion in the pan for 4 minutes. Stir in the wine. Blend together the sour cream and egg yolks and add to the pan. Simmer until slightly thickened and season with salt and pepper, if desired.

Strain into the serving dish and stir in the remaining pimientos, diced.

Spoon some sauce over each breast and serve. Pass the remaining sauce in a bowl at the table for those who desire more.

Peppers

Pepper Scramble
4 servings

1 small hot chili pepper
2 small green peppers
1 tablespoon olive oil
½ cup chopped green onion
6 eggs
½ cup sour cream
1 medium tomato
Pepper, to taste
3 tablespoons butter
4 slices toast
Paprika

Seed and finely chop the chili pepper.
Seed and chop one green pepper and seed
and slice the other. Set the slices aside.

Heat the oil in a skillet or wok and
stir-fry the chopped green and chili
peppers and onion until soft.

Remove and set aside to cool.

Beat the eggs with ¼ cup of the sour
cream. Peel and seed the tomato and cut
into eighths. Add the sautéed vegetables
and tomato pieces to the egg mixture and
season with the pepper.

Melt the butter in the wok or skillet and
cook the egg mixture over medium heat
until scrambled.

Place a toast on each plate. With an ice
cream scoop, scoop a mound of egg onto
the center of each toast. Top with a dollop
of the remaining sour cream and sprinkle
with paprika. Garnish each plate with 2
green pepper slices and serve.

Green Bean Pepper Crunch
8 to 10 servings

1 can (15½-ounce) French-style
  green beans
1 package (3-ounce) lemon flavored
  gelatin
1 envelope unflavored gelatin
1 green pepper, chopped
½ cup onion, minced
½ cup celery, chopped
½ cup nuts, chopped

Sauce:
½ cup cucumber, grated and drained
1 cup mayonnaise
½ cup green pepper, finely chopped
2 teaspoons vinegar
½ teaspoon salt or ¼ teaspoon dill
Freshly ground black pepper

Drain the liquid from the green beans. Add
enough water to it to make 2 cups liquid.
Bring to a boil and dissolve the lemon
gelatin in the hot liquid. Dissolve the
unflavored gelatin in ¼ cup cold water and
add to the lemon gelatin mixture. Cool until
slightly thickened.

Stir in the remaining ingredients. Pour
into individual molds and chill until firm.

To make the sauce, stir all the
ingredients together and chill.

To serve, unmold the salad onto crisp
greens and top with the sauce.

The white potato, or *pomme de terre*, is of the nightshade family and is grown or sprouted from a chunk of itself that contains one or two eyes, or buds. Potato flowers can be white, pink, yellow or blue. The potato is valuable because of its starch content. Choose firm, clean and relatively smooth potatoes which are free of cuts or bruises. Avoid green potatoes or those with sprouts.

Delmonico Potatoes

4 servings

2 tablespoons butter
4 cups boiled potatoes
4 hard-cooked eggs
Cream Sauce*
Salt and pepper (optional)
¾ cup grated medium Cheddar cheese
½ cup fine bread crumbs

Butter a baking or casserole dish with one tablespoon of the butter. Slice the potatoes and the eggs. Make the Cream Sauce*, substituting ½ pint of cream for the one cup of milk. Season with salt and pepper, if desired.

Alternate layers of boiled potatoes and egg slices in the baking dish. Pour the Cream Sauce over and sprinkle the surface with cheese and the bread crumbs. Dot with the remaining butter and bake 30 to 40 minutes at 350 degrees F. until golden brown. Serve from the baking dish.

Potatoes

German Potato Salad

6 to 8 servings

6-8 large potatoes
3 tablespoons bacon fat
1 large onion, chopped
½ cup chopped celery
1 cup heavy cream
½ cup cider vinegar
½ teaspoon dry mustard
2 teaspoons salt or 1 teaspoon chopped
   basil plus 1 clove garlic, minced
½ teaspoon pepper
Salad greens

Wash the potatoes well and boil them in
their jackets about 40 minutes or until
tender. Meanwhile, gently fry the onion and
celery in the bacon fat just until tender and
heat the cream with the vinegar, mustard,
salt and pepper. Do not boil.

Drain the potatoes. Peel them while they
are hot and slice them into a wooden salad
bowl. Mix well with the onion mixture and
the dressing. Serve hot with a ring of salad
greens around the edge of the bowl.

Eat lukewarm—never cold.

Crock Pot Baked Potatoes

4 servings

4 large potatoes
5 tablespoons butter
1 cup sour cream
1 jar (8-ounce) processed cheese
   spread
½ pound shaved cooked ham
½ pound shaved cooked roast beef
Bacon bits

Wash the potatoes, pierce them with a fork
and grease well with one tablespoon of the
butter. Put in the crock pot and cook on
low for 8 hours.

At mealtime, open each potato wide
and put one tablespoon butter and
one tablespoon sour cream into each.
Sprinkle the shaved ham and roast beef
generously over all, top with one
tablespoon of the cheese and bacon bits
and serve.

Vichyssoise

6 servings

3 medium potatoes
1 medium onion, chopped
1½ cups chicken broth
¾ cup heavy cream
½ teaspoon salt (optional)
¼ teaspoon white pepper
¾ cup crushed ice
Fresh parsley

Wash the potatoes well and boil them in their jackets about 40 minutes or until tender. Drain. Peel them while hot and dice to get about 1½ cups cooked potatoes.

Place the potatoes, onion, broth, cream and seasonings in the blender and spin for 15 seconds. Add in the crushed ice and spin for 15 seconds more.

Serve immediately in individual soup bowls, garnished with a sprig of fresh parsley.

Potato Pancakes

6 to 8 servings

6 medium potatoes
4 eggs
½ cup flour
1 large onion, chopped
1 teaspoon salt (optional)
¼ teaspoon white pepper
½ teaspoon baking powder
1 tablespoon butter

Wash and peel the potatoes and cut into cubes. Do not cook.

Place the raw potato cubes, eggs, flour, onion, salt and baking powder in the blender. Spin for 10 seconds. Stop the motor, stir, and spin again until smooth and creamy, about 10 seconds.

Melt the butter on a griddle and ladle the pancake batter (about ¼ cup at a time) onto the hot griddle. When golden brown on one side, flip and brown the other side.

Serve with a pitcher of honey and plenty of butter.

Potatoes

Q's, R's, S's and T's

The radish is a root vegetable of the mustard family. It is a biennial, taking two years to produce its seeds. Radishes are usually pulled up the first year, however, for food, before they grow their flower stalks. The most common radish is the spring variety which is small, round and white with a thin red skin. Radishes are almost always eaten raw. Select those that are smooth, firm and well shaped with few cuts or black spots.

Gardentime Salad

4 servings

½ cup sliced radishes
1 can (8½-ounce) diagonal-cut green
 beans, drained
1 can (8½-ounce) baby peas, drained
½ cucumber, sliced
12 cherry tomatoes, halved
⅓ cup Italian dressing
4 hard-cooked eggs, cut in wedges

Combine the vegetables in a large salad bowl and toss. Pour the dressing over all.

Cover and chill. Allow to marinate in the refrigerator for at least 2 hours.

Garnish the bowl with the egg wedges at serving time.

Radish Rolls

12 to 15 servings

½ **pound radishes**
⅔ **cup minced walnuts**
1 package (8-ounce) cream cheese,
  softened
⅓ **cup crumbled blue cheese**
1 can (4½-ounce) deviled ham
1 tablespoon grated onion
¼ **teaspoon curry powder**
Worcestershire sauce, to taste
Carrot curls
Fresh parsley

Wash the radishes well and cut off the ends. Thinly slice two, place in a jar of cold water, cover and refrigerate. Grate the rest and mix in a shallow bowl with the walnuts. Set aside.

Combine the cheeses, ham, onion, curry and Worcestershire sauce. Mix until smooth and well blended.

Form into small logs, about ½x1½ inches. Roll in the radish/walnut mixture and place on a sheet of wax paper.

Wrap tightly and refrigerate until well chilled.

Arrange on a medium platter and garnish with the drained radish slices, carrot curls and sprigs of fresh parsley. Serve with an assortment of crackers and party breads.

Radishes

Rose Radishes in Cream

4 servings

1 pound radishes
4 tablespoons butter
1½ cups whipping cream
Black pepper

Peel the radishes and parboil them in water for 6 minutes. Drain well.

Heat the butter over moderate heat and cook the radishes in the butter, tossing frequently so that they are evenly coated.

Add the cream and cook down to approximately one cup of cream.

Sprinkle with the black pepper and serve hot.

Rhubarb grows in red or green stalks which are 18 to 36 inches tall and are topped with a 12-inch span of poisonous leaves. Select bright-colored, crisp and firm stalks of medium size. Oversized stalks may be tough and stringy.

Rhubarb Cobbler

9 servings

3½ cups frozen cut rhubarb
1 package (3-ounce) strawberry flavored
   gelatin
1¼ cups biscuit mix
½ cup plus 2 tablespoons sugar
¼ cup milk
1 egg, beaten
½ cup flour
¼ cup butter
Plain yogurt

Grease an 8x8x2-inch baking pan.
Combine the rhubarb and gelatin, mix well
and set aside.
   Combine the biscuit mix, 2 tablespoons
sugar, milk and egg. Blend well. Spread
the dough over the bottom of the prepared
pan. Spread the rhubarb mixture over top
of the dough.
   Combine the remaining sugar and flour
and cut in the butter until the mixture
resembles fine crumbs. Sprinkle over the
rhubarb. Bake for 35 to 40 minutes at 375
degrees F.
   Serve warm, topped with a dollop of
fresh plain yogurt.

Rhubarb Walnut Bread

2 small loaves

1½ cups light brown sugar
⅔ cup safflower oil
1 egg
1 cup buttermilk
1 teaspoon baking soda
½ teaspoon salt or salt substitute
1 teaspoon vanilla
2½ cups flour
1½ cups diced raw rhubarb
½ cup chopped walnuts
1 tablespoon melted butter
5 tablespoons sugar

Butter two small loaf tins and dust them
with flour.
   Beat the brown sugar, oil and egg in an
electric mixing bowl until well blended.
   Combine the buttermilk, baking soda,
salt and vanilla in a 2-cup measuring
pitcher. Alternately add the milk mixture
and the flour to the sugar mixture, beating
continuously.
   Reduce the beater to slow and fold in
the rhubarb and the nuts.
   Divide the batter between the prepared
pans. Mix the butter and sugar and spread
over the top of the loaves.
   Bake 40 to 45 minutes at 325 degrees F.
or until an inserted knife comes out clean.
Remove from the pan.
   Serve sliced and spread with cream
cheese.

Rhubarb

The rutabaga, closely related to the turnip, is also of the cabbage family. The edible part is its yellow, fleshy root, giving it the name yellow turnip. It is also known as the cow turnip, turnip-rooted cabbage or Swedish turnip. Rutabagas have a more pungent flavor than the true white turnips, which are dealt with later in this section. Choose rutabagas that are smooth, firm and heavy for their size, but not large.

Mashed Rutabagas

6 servings

1 (1-pound) rutabaga
2 cups cooked, diced potatoes
4 tablespoons butter
2 tablespoons brown sugar
2 tablespoons dry sherry (optional)
1½ teaspoons salt or 1 clove garlic, minced
¼ teaspoon caraway seeds
¼ teaspoon pepper
Chopped parsley

Peel the rutabaga and cut into chunks. Cover with boiling water. Cover and cook for 25 to 35 minutes or until tender. Drain and return to the pan. Toss over low heat until the steam stops rising from the vegetable. Keep shaking the pan so that the rutabaga will not be scorched.

In a blender, mash or puree the cooked rutabaga along with the potatoes. Add the butter, sugar and sherry and spin again. Return the mixture to the pan and stir over moderate heat until quite dry.

Add the seasonings and serve hot sprinkled with chopped parsley.

Yellow Turnip Soufflé

4 servings

1 (1-pound) rutabaga
2 tablespoons butter
2 tablespoons flour
1 cup cream
2 tablespoons chopped scallions
3 egg yolks
½ teaspoon cloves
½ teaspoon nutmeg
Salt and pepper (optional)
3 egg whites

Peel the rutabaga and cut it into cubes. Boil it in a little water until tender. Drain and force through a food mill or puree it in a food blender. Return to the pan and cook until thick and almost dry.

Heat the butter in a small saucepan and stir in the flour. When blended add the cream and whisk until smooth. Remove from the heat.

Add the scallions and the rutabaga puree.

Beat the egg yolks with the cream sauce and add the cloves and nutmeg. Add this mixture gradually to the rutabaga mixture, stirring continually. Season to taste with salt and pepper. Cool.

About 45 minutes before serving, preheat the oven to 350 degrees F. Beat the egg whites until stiff and fold them into the rutabaga base. Pour into a 1-quart buttered soufflé dish. Bake 35 to 40 minutes.

Rutabagas

Germany lays special claim to the sour cabbage or sauerkraut, which is made from pressed layers of salt and shredded cabbage. Its taste and texture is so far removed from that of fresh cabbage, that we have given it its own space in this book. Choose the bulk variety for its crispness and flavor.

Sauerkraut in Gin and Sour Cream

4 to 6 servings

2-3 pounds bulk sauerkraut
½ cup boiling water
1 bay leaf
½ teaspoon thyme
½ teaspoon white pepper
½ cup gin
1 tablespoon butter
4 tablespoons chopped onion
2 cooking apples
½ pint sour cream

Wash the sauerkraut in several waters. Drain and soak in fresh cold water for 20 minutes. Drain. Place in a heatproof serving dish. Add the boiling water, herbs, pepper and gin. Cook gently, partially covered, for 30 minutes. Stir occasionally.

Meanwhile heat the butter and sauté the onions until golden brown.

Peel, quarter and core the apples. Cut into small cubes and add to the onion.

Cook 2 minutes and add the sour cream. Stir until the cream is heated but not boiling.

Remove the bay leaf from the sauerkraut and stir in the cream mixture. Mix well before serving.

Sauerkraut Shepherd's Pie

6 servings

3 pounds bulk sauerkraut
¼ pound salt pork
2 onions, chopped
6 knockwurst
½ cup boiling water
¼ teaspoon coarse black pepper
1 teaspoon caraway seeds
5 medium potatoes
½ cup hot cream
3 tablespoons butter
1 egg, slightly beaten
Salt and pepper (optional)
Parsley

Wash the sauerkraut in several waters. Squeeze out the water and then soak in cold water for 10 to 20 minutes. Squeeze dry again or fluff in a lettuce drier or colander.

Dice the salt pork in ½-inch cubes and sauté in a skillet over medium heat until the pork renders two tablespoons of liquid fat. Add the onions and continue cooking, stirring occasionally, until the onion is soft and golden brown. Remove from the heat.

Slice the knockwurst in ¼-inch slices.

Spread half the sauerkraut in the bottom of a well-buttered 2-quart casserole. Cover with half the salt pork and onion mixture with the fat, and half the knockwurst. Sprinkle with half the pepper and caraway seeds. Cover with the remaining sauerkraut and repeat the process.

Meanwhile boil the potatoes in 4 cups of water for 25 minutes or until tender. Drain and toss over moderate heat in the saucepan until the potatoes become mealy on the outside. Remove from the heat and mash well. Stir in the cream, two tablespoons of the butter and the egg, beating vigorously until smooth. Season with salt and pepper, if desired.

Spread the potato over the top layer of the casserole and brush with the remaining tablespoon of butter. Bake at 400 degrees F. until the top is golden brown. Garnish with sprigs of parsley.

Sauerkraut

Pineapple Sauerkraut Salad

4 servings

1½ pounds bulk sauerkraut
3 scallions or green onions, chopped
3 tablespoons peanut, safflower or
corn oil
1 small can (8-ounce) pineapple bits
packed in juice
Freshly ground black pepper

Wash the sauerkraut in several waters.
Drain thoroughly and dry. Chop the
sauerkraut quite fine.

Add the diced scallions and oil and toss.

Add the pineapple bits and one
tablespoon of juice and several grinds of
black pepper. Toss again and chill in the
refrigerator. Toss just before serving.

Snow peas or sugar peas are a variety of
pea which is eaten pod and all and is
frequently used in Oriental cooking. The
pod is flat and pale green and the peas
inside are tiny and underdeveloped. Select
pods no longer than 3 inches containing
peas that are barely formed.

Chinese Pork with Snow Peas

6 servings

6 pork chops, ½ inch thick
1 teaspoon sesame oil
2 tablespoons soy sauce or tamari
1 tablespoon cornstarch
1 can (16-ounce) peach slices
¾ cup plus 1 tablespoon safflower oil
1 package (10-ounce) frozen snow peas
1 teaspoon cornstarch

Trim the fat and bones from the meat and discard. Put the meat between sheets of wax paper. Pound with a mallet till ¼ inch thick. Cut into bite-size cubes and place in a shallow bowl. Combine the sesame oil, soy sauce and one tablespoon cornstarch in a bowl and pour over the pork. Allow to marinate for 15 minutes.

Drain the peaches, reserving ¼ cup of the liquid and chop into bite-size chunks.

In a skillet or wok, heat the ¾ cup oil. Stir-fry the pork cubes until browned on all sides. Remove and set aside.

Run the pods under hot water to thaw. Pat dry. Heat the one tablespoon oil in the skillet or wok and add the snow peas. Add water, cover and steam for 30 seconds.

Add the peach chunks and pork to the snow peas and stir-fry until heated through. Dissolve the one teaspoon cornstarch in the reserved peach syrup and stir into the contents of the skillet or wok. Cook over moderate heat, stirring until sauce becomes thick. Serve over hot, fluffy rice.

Snow Pea Stir-fry

6 to 8 servings

1 package (10-ounce) frozen snow peas
4 boneless chicken breasts
2 tablespoons cooking sherry
3 tablespoons soy sauce or tamari
1 tablespoon cornstarch
1 teaspoon ground ginger
⅛ teaspoon crushed red pepper
3 medium zucchini
1 pound fresh mushrooms
½ cup olive oil

Run the frozen snow peas under hot water to thaw and pat dry with paper toweling.

Cut the chicken breasts into bite-size pieces. Mix the sherry, soy sauce, cornstarch, ginger and red pepper in a bowl and stir in the chicken pieces. Set aside.

Wash the zucchini, cut off the ends and slice. Wipe the mushrooms clean.

Heat the oil in a skillet or wok and stir-fry the zucchini slices and whole mushrooms. When the zucchini is browned, remove both the zucchini and mushrooms to paper toweling.

Stir-fry the chicken until tender in the oil remaining in the skillet or wok. Stir in the snow peas and add the zucchini and mushrooms back in. Heat through.

Spoon onto mounds of hot, fluffy rice and serve.

Snow Peas

All squash are of the gourd family and have a net of seeds in the center which forms a hollow when the vegetable is ripe. Winter, banana, butternut, green and blue hubbard, spaghetti and acorn squash are among those classified as winter squash. Zucchini and other summer squash are dealt with in the last section of this book. When selecting fresh squash, avoid specimens with any soft areas.

Stuffed Acorn Squash

4 to 6 servings

3-4 acorn squash
1 cup boiling water
Butter
Salt and pepper (optional)
3 apples, sliced
2 teaspoons brown sugar
Nutmeg

Halve the squash, remove the seeds and fibers and place the halves cut side down in a baking dish. Pour the boiling water over the halves and bake 30 minutes at 350 degrees F.

Turn the halves over, dot each with a little butter and sprinkle with salt and pepper, if desired.

Fill the cavities with the apple slices and dot with more butter. Sprinkle each with the brown sugar and a little nutmeg.

Return to the oven and bake until the squash is tender, about 30 minutes.

Serve with hot cider and cinnamon sticks.

Winter Squash with Applesauce

4 servings

2 winter squash
4 tablespoons butter
Salt and pepper (optional)
⅔ cup applesauce
2 teaspoons horseradish
Cinnamon

Prepare the squash as you would for Stuffed Acorn Squash*, baking for 50 minutes instead of 30.

In a small saucepan, mix together the applesauce and horseradish. Heat through.

Put a large spoonful of the warmed applesauce mixture into each squash half, sprinkle with cinnamon and serve.

Winter Squash Bread

1 loaf

1 acorn or butternut squash
¼ cup honey
½ cup corn oil
1½ cups flour
¼ teaspoon baking powder
½ teaspoon soda
1½ teaspoons cinnamon
½ cup chopped walnuts

Cut the squash in quarters and remove the seeds and fibers. Pare the pieces and chop them into cubes to equal one cup. Drop them into a pan of boiling water and cook for 10 to 12 minutes or until tender. Drain thoroughly. Mash the squash with a fork.

In a large mixing bowl, combine the squash, honey and oil. The mixture will be runny.

Blend in ½ cup of the flour, the leavenings and cinnamon. Add the rest of the flour, ½ cup at a time, adding the nuts with the last ½ cup.

Pat the mixture into the bottom of a well-greased loaf pan. Bake at 325 degrees F. for one hour or until a knife inserted comes out clean.

Cool on a rack before slicing.

Squash, winter

Winter Squash with Bacon

4 to 6 servings

3-4 pounds butternut squash
½ pound bacon
4-6 tablespoons butter
Salt and pepper (optional)

Cut the squash in quarters and remove the seeds and fibers. Pare the pieces and drop them into a pan of boiling water. Boil for 20 to 25 minutes or until tender. Drain thoroughly.

Meanwhile fry the bacon until crisp in a moderately heated skillet. Drain the bacon on toweling and keep warm.

Mash the squash with a potato masher or with an electric beater. If the squash is too liquid stir it over heat until the excess moisture evaporates. Season with butter and salt and pepper and serve in a heated vegetable dish with the bacon crumbled over the top.

The sixteenth-century French called it a *pomme d'amour,* or "love apple," and today it is one of our most versatile and popular vegetables. What started as a berry filled with seeds has been cultivated to produce a variety in colors, sizes and shapes, from the small, red cherry tomato, to the juicy yellow, to the large steak or even red or yellow pear-shaped tomato. Choose smooth, firm and plump-looking tomatoes with good color.

Iced Tomato Soup

3 to 4 servings

2 large tomatoes
2 tablespoons chopped green onion
2 tablespoons chopped green pepper
¼ cucumber, sliced
1 clove garlic, pressed
1 teaspoon salt or 2 teaspoons chopped
   fresh or 1 teaspoon dried basil
Pepper, to taste
1 tablespoon olive oil
1 tablespoon red wine vinegar
¼ cup red wine
Worcestershire sauce, to taste
½ cup crushed ice
Plain yogurt
Parsley

Place all the ingredients but the ice, yogurt and parsley in the blender. Spin once, but do not overblend.

Add the crushed ice, spin again.

Pour into individual soup bowls and serve cold, garnished with a dollop of yogurt and a sprig of parsley.

Baked Stuffed Tomatoes

6 to 8 servings

6-8 medium tomatoes
5 tablespoons butter
2 tablespoons minced onion
½ cup finely diced celery
1½ cups soft white bread crumbs
Salt and pepper (optional)
Sugar
¼ cup cracker crumbs
¼ cup grated Parmesan cheese

Cut a thin slice from the top of each tomato and scoop out the pulp, leaving a ½-inch shell. Force the pulp through a sieve to remove the seeds. Turn the tomato shells upside down on a rack to drain.

Heat 3 tablespoons of the butter in a small pan and sauté the onion and celery just until the onion is soft. Add the bread crumbs and tomato pulp and stir until well mixed.

Sprinkle the inside of the tomatoes with a little salt and pepper, if desired, and unless the tomatoes are garden fresh, add just a trace of sugar. Fill with the stuffing.

Mix together the cracker crumbs and Parmesan cheese and sprinkle over top of the tomatoes. Dot with the remaining butter and bake for 20 minutes at 375 degrees F.

Tomatoes

Stuffed Tomato Flowerets

4 servings

1½ cups cooked and diced chicken
½ pound Cheddar cheese, cubed
1 can (11-ounce) mandarin oranges,
  drained
2 bananas, chopped
½ cup slivered almonds
½ cup mayonnaise
4-6 medium tomatoes
4 lettuce leaves
1 hard-cooked egg, sliced

Combine the first six ingredients in a medium bowl. Toss gently to blend and set aside.

Wash and peel the tomatoes and cut out the stems. Slice them with a sharp knife into quarters, leaving them attached at the bottom.

Place a lettuce leaf on each salad plate. Put one tomato floweret in the center of each. Fill with the chicken salad and garnish with the egg slices.

Green Tomatoes in Cream Sauce

6 to 8 servings

3 large green tomatoes
1 cup flour
1 teaspoon salt or ½ teaspoon crushed
  red pepper
½ teaspoon dried basil
¼ teaspoon pepper
½ teaspoon sugar
12-18 slices bacon
Cream Sauce*

Cut the tomatoes into thick slices, allowing 2 to 3 slices to a serving. Combine the flour, seasonings and sugar. Spread on a plate.

Fry the bacon in a skillet until crisp and transfer to paper toweling to drain.

Make the Cream Sauce*. Dip each tomato slice in the flour mixture and fry in the bacon fat 2 to 3 minutes on each side or until golden brown. Place on a piece of johnnycake or cornbread.

Cover with a little Cream Sauce and top with a crisscross of bacon slices.

The true turnip is a small white tuber topped with a purple crown. It is of the cabbage family, but has a more delicate flavor than its relative, the yellow turnip or rutabaga (dealt with earlier in this section). Turnips should be firm, smooth and of medium size.

Glazed Turnips

4 servings

6-8 turnips
4 cups chicken broth
2 tablespoons sugar
3-4 tablespoons butter
Salt and pepper (optional)
2 tablespoons sherry
¾ teaspoon salt or ½ teaspoon allspice
⅛ teaspoon white pepper
⅛ teaspoon nutmeg

Pare and slice the turnips into ¼-inch slices. Cook them in the chicken broth in a covered pan for 10 minutes or until almost tender.

Remove the cover and add the remaining ingredients. Boil down slowly, uncovered, until the turnips are glazed. Keep an eye on them to see that they do not burn. Toss occasionally.

These can be cooked in advance, transferred to a baking dish and reheated in the oven.

Turnips

Scalloped Turnips

6 servings

4 cups peeled and sliced turnips
4 tablespoons minced onion
4 tablespoons crumbled cooked bacon
3 tablespoons flour
6 tablespoons butter
Salt and pepper (optional)
1½ cups hot milk
Paprika

Cover the sliced turnips with boiling water and cook for 3 minutes.

Drain and rinse in cold water.

Butter a deep baking dish and cover the bottom with a layer of turnips. Sprinkle with minced onion, bacon and flour. Dot with butter. Sprinkle with salt and pepper, if desired. Repeat the process until the dish is filled.

Pour the hot milk over the turnips. They should be just covered.

Bake 45 minutes at 350 degrees F. Garnish with a dash of paprika before serving.

Creamed Turnips

4 servings

2-3 pounds turnips
4 tablespoons butter
Salt and pepper (optional)
2 tablespoons sherry
¼ teaspoon freshly grated nutmeg
1 cup cream
Parsley

Peel and dice the turnips into ¾-inch cubes. Boil in water for 10 minutes or just until tender. Drain and toss in the butter and season with salt and pepper, if desired.

Heat the sherry, nutmeg and cream in a saucepan, but do not boil. Pour over the buttered turnips, garnish with fresh parsley sprigs and serve.

Turnip Soup

6 servings

1 pound turnips
1 onion
2 pints chicken broth
1 clove
1 cup water
3 tablespoons butter
2 tablespoons flour
2 egg yolks
½ cup cream
Salt and pepper (optional)
Chopped parsley

Peel the turnips and onion and slice coarsely. Place in a saucepan with the chicken broth, clove and water. Cover partially and cook until the vegetables are tender. Strain the liquid into the top part of a double boiler. Discard the clove.

Spin the vegetables with one cup of broth in a blender or food processor.

Add the butter and flour and continue to blend for a few seconds. Add the mixture to the vegetable broth. Reheat over boiling water.

Beat the egg yolks and cream until blended. Add one cup of the hot broth gradually and when well mixed stir into the soup. Add salt and pepper, if desired.

Continue to stir until the soup is hot and thickened. Pour into individual soup cups and garnish each with the chopped parsley. Serve with an assortment of crackers and melba toasts.

Turnips

U's, V's, W's and X's

One appealing characteristic of vegetables is that, though delicious individually, they are scrumptious when combined. The following recipes use different vegetables in various combinations, resulting in a unique taste experience.

Vegetable Mayonnaise

4 to 6 servings

2 cups cold cooked cauliflower
2 cups cold boiled cubed potatoes
1 cup cold diced cooked carrots
1 cup cold cooked peas
Salt and pepper (optional)
1 cup mayonnaise
Lettuce cups
½ cup sour cream
3 tablespoons chopped chives

Toss the vegetables together, sprinkling them with a little salt and pepper, if desired. Bind them together with ¼ cup of the mayonnaise.

Fill the lettuce cups with the mixture and place them on a serving platter.

Mix the rest of the mayonnaise with the sour cream. Cover the vegetables with the mixture and sprinkle liberally with the chopped chives.

Vegetable Juice

4 servings

4 cups tomato juice
2 cups chopped celery
1½ cups chopped carrot
1 large onion, chopped
4 tablespoons lemon juice
Worcestershire sauce, to taste
Hot pepper sauce, to taste
⅓ cup sour cream
Chopped parsley

Place the first four ingredients in the blender and spin for 15 seconds. Add the lemon juice and sauces. Spin again and taste for seasoning.

Pour into four good-sized glasses, top each with a dollop of sour cream and sprinkle with chopped parsley.

Vegetable Shish Kabob

6 servings

2-2½ pounds of lamb
½ cup olive oil
¼ teaspoon curry powder
2 teaspoons oregano
1 teaspoon salt (optional)
½ teaspoon pepper
3 onions, minced
4 large tomatoes
2 green peppers
1 cup mushroom caps

Cut the meat from the leg of lamb into bite-size pieces. Place in a medium bowl. Add to the lamb ¼ cup of the oil, the curry, oregano, salt, pepper and onion. Cover and refrigerate for several hours, preferably overnight.

Cut the tomatoes into wedges and the peppers into square pieces. Leave the mushrooms caps whole.

On 6 skewers, alternate the lamb chunks, tomato wedges, peppers and mushrooms. Brush with the remaining oil and grill for 30 minutes, turning to cook evenly.

Vegetable Combinations

Vegetable-Cheese Soup

5 servings

1 package (10-ounce) mixed vegetables
  frozen in butter sauce
¼ cup butter
½ cup chopped onion
½ cup flour
⅛ teaspoon pepper
1¾ cups milk
1 can (10¾-ounce) chicken broth
1 jar (8-ounce) processed cheese spread
1½ teaspoons prepared mustard

Cook the vegetables according to the
package directions. Meanwhile, in a
medium saucepan, sauté the onion in the
butter.

Blend in the flour, salt and pepper. Stir
in the milk gradually.

Heat to boiling, stirring constantly.
Reduce the heat and add the remaining
ingredients and the vegetables.

Heat thoroughly and serve.

Vegetable Medley

10 servings

1 package (10-ounce) sweet peas frozen
  in butter sauce
1 package (10-ounce) cut broccoli frozen
  in cheese sauce
1 package (10-ounce) whole kernel corn
  frozen in butter sauce
2 tablespoons chopped pimiento
⅓ cup sliced almonds
1 chicken bouillon cube
½ cup hot water
½ teaspoon instant minced onion
¼ teaspoon dry mustard

Prepare the peas, broccoli and corn
according to the package directions.
Combine them in a medium saucepan and
add the pimiento and almonds.

In a small bowl, dissolve the bouillon in
the hot water and add the onion and dry
mustard.

Stir the bouillon mixture into the
vegetables. Heat through and serve.

The water chestnut is not a true vegetable, but the fruit of a water plant. Its crunchy, watery texture enhances many Oriental dishes and is becoming more and more popular in American cooking. Select smooth, firm water chestnuts of small size.

Green Beans and Water Chestnuts

6 servings

**1 pound fresh green beans or
  2 packages (10-ounce) frozen cut-up
  green beans
1 can (8-ounce) water chestnuts,
  drained
1 can (10-ounce) cream of mushroom
  soup
1 cup grated American cheese
½ cup slivered almonds
Salt and pepper (optional)**

Wash the fresh green beans and cut them up. If using frozen, run them under hot water until separated. Thinly slice the water chestnuts.

In a well-greased casserole dish, combine the mushroom soup, cheese, cut-up beans, sliced water chestnuts, almonds and seasoning.

Reserve some grated cheese and slivered almonds to sprinkle on top, if desired.

Bake at 350 degrees F. for 30 minutes, or until the beans are tender and the cheese is hot and bubbly.

Water Chestnuts

Water Chestnut Treasure Chest

8 to 10 servings

2 packages (11-ounce) frozen rice pilaf
2 packages (10-ounce) cut broccoli
   frozen in cheese sauce
2 cups cut-up cooked turkey
1 can (8-ounce) water chestnuts,
   drained
1 can (10-ounce) cream of mushroom
   soup
4 small mushrooms, sliced
½ cup mayonnaise
1 tablespoon butter, melted
½ cup herb croutons

Cook the rice and broccoli according to
the package directions. Spread the rice in
the bottom of a shallow baking dish. Cover
with the turkey and then the broccoli in
cheese sauce.

   Slice the water chestnuts and combine
them in a small bowl with the soup,
mushrooms and mayonnaise. Pour this
mixture over the broccoli.

   Toss the croutons in the butter and
sprinkle them over top of the casserole.

   Bake at 350 degrees F. for 40 to 50
minutes.

Shrimp and Water Chestnut Crêpes

6 to 8 servings

6-8 crêpes
1 can (8-ounce) water chestnuts, drained
2 tablespoons butter
2 tablespoons chopped green onion
1 can (10-ounce) cream of shrimp soup
1 cup cooked baby shrimp
¼ cup milk
1 tablespoon sherry
Worcestershire sauce, to taste
Soy sauce or tamari, to taste
Cream Sauce*
2 cups shredded Swiss or Gruyère
   cheese
Paprika
Parsley

Prepare the crêpes with a crêpe-maker
and set aside.

   Slice the water chestnuts. Melt the
butter in a skillet and sauté the water
chestnut slices and onion until tender, but
still crisp. Stir in the remaining ingredients
down to but *not* including the Cream
Sauce*. Heat through, stirring frequently.

   Meanwhile, make the Cream Sauce
and stir in the shredded cheese until
melted.

   Put a large spoonful of shrimp mixture in
the center of each crêpe and fold into the
traditional roll. Dribble a generous amount
of the cheese sauce over each crêpe,
sprinkle with paprika and serve garnished
with fresh parsley sprigs.

Watercress grows in cold, freshwater streams. Its fleshy leaves taste somewhat like pepper and are used as greens, raw or cooked. Select fresh, green and tender leaves with no bruises or brown spots.

Healthful Watercress Munchies

20 servings

1 loaf homemade bread
½ pound butter, softened
1 pound watercress
½ cup mayonnaise
½ cup French Dressing*

Cut off the heel of the loaf of bread. Butter the cut end of the loaf and cut off as thin a slice as possible. Continue until you have as many slices as needed (2 per serving). Place all the slices buttered side up.

Wash and dry two sprigs of watercress for each sandwich. Combine the mayonnaise and French Dressing* in a bowl. Dip the sprigs into the dressing mixture and place two in the corner of each buttered slice of bread. Place the leafy tips toward the outside. Roll each slice diagonally so that the leaves just show at the end. Secure with party toothpicks.

Cover and refrigerate until time to serve.

Watercress

Cream of Watercress Soup

4 servings

1 bunch watercress
2 tablespoons butter
1 small onion, chopped
½ teaspoon curry powder
2 potatoes
2 cups chicken broth
Salt and pepper (optional)
½ cup cream

The following recipes of sauces and dressings can add that "something extra" to vegetable dishes or to any favorite recipe. All are listed as ingredients in recipes throughout this book. Check the Index for suggested combinations.

Snip a half inch off the watercress stems and discard. Wash the watercress thoroughly. Cut off the stems and chop into short pieces. Put the leaves aside.

Melt the butter in a skillet and sauté the onion until tender. Add the curry powder and stir until the curry is browned. Toss in the watercress stem pieces and cook for 5 minutes, stirring frequently.

Peel the potatoes and chop them into the skillet. Pour in the chicken broth and simmer until the potatoes are tender.

Pour the mixture into the blender and spin for 15 seconds. Next, pour the mixture back into the skillet through a strainer (to remove the watercress stems). Add the watercress leaves and cook over medium heat for one minute.

Spin once more in the blender and season with salt and pepper, if desired.

Return the soup to the skillet and stir in the cream. Heat through, but do not boil. Serve with wheat melba toast.

Vinaigrette

1½ cups

⅓ cup vinegar (red wine, white wine,
    tarragon or cider)
1 cup salad oil or ½ cup olive oil plus
    ½ cup safflower, peanut or corn oil
¾ teaspoon salt or salt substitute
½ teaspoon freshly ground black pepper

Place the ingredients in a small jar and
shake well before using.

Blender Method Mayonnaise

1 cup

1 egg
2 tablespoons lemon juice
1 cup salad oil or ½ cup olive oil
    plus ½ cup corn, peanut or
    safflower oil
½ teaspoon salt or salt substitute
½ teaspoon pepper
½ teaspoon dry mustard
1 tablespoon boiling water

Put the egg, lemon juice, 4 tablespoons of
the oil, the salt, pepper and mustard in the
blender. Cover and spin while you count to
six slowly.

Remove the center of the cover and
gradually add the rest of the oil, pouring it
in a thin, steady stream while continuing to
spin.

Add the water and stop spinning.

Transfer to a covered jar and store in
the refrigerator.

X-tra Sauces

Beater Method Mayonnaise

2 cups

3 egg yolks
1 teaspoon salt or salt substitute
1 teaspoon Dijon mustard
½ teaspoon white pepper
2 cups salad oil or 1 cup olive oil
**   plus 1 cup corn, peanut or**
**   safflower oil**
2 tablespoons wine or cider vinegar
2 tablespoons boiling water

Combine the egg yolks, salt, mustard and pepper in the bowl of an electric mixer and beat until thick.

Start adding the oil in a thin stream without stopping the beater. When ¾ cup of oil has been added, add the vinegar slowly and continue with the rest of the oil. Beat in the water.

Transfer to a covered jar and store in the refrigerator.

Cream Sauce

2 cups

4 tablespoons butter
4 tablespoons flour
2 cups cold milk
Salt and pepper (optional)

Heat the butter in a saucepan. Stir in the flour and cook the mixture 2 minutes, stirring with a whisk.

Add half the milk and stir hard while it comes to a boil. When the sauce is smooth, add the remaining milk, still stirring. Stir until the sauce is thick and smooth. Season with salt and pepper, if desired.

This can be doubled or halved. It can be stored in the refrigerator in a covered container for several days or in the freezer indefinitely.

French Dressing

1 cup

2 teaspoons salt or salt substitute
½ teaspoon freshly ground black pepper
¼ to ⅓ cup vinegar (wine, cider or
    tarragon)
1 cup olive oil or 1 cup peanut or
    vegetable oil or ½ cup each
¼ teaspoon powdered Vitamin C

Put all the ingredients except the Vitamin C
into a small covered jar and shake well.
The dressing can be stored in the
refrigerator. Add the Vitamin C and shake
again just before serving.

Quick Tomato Sauce

1½ cups

2 tablespoons butter
2 tablespoons olive oil
½ cup chopped onion
1 large clove garlic, pressed
1 can (20-ounce) plum tomatoes,
    drained
1 can (6-ounce) tomato paste
½ cup red wine
1 cup water
½ teaspoon sugar
¼ teaspoon black pepper
2 teaspoons chopped fresh basil or
    ½ teaspoon dried basil
½ teaspoon oregano

In a saucepan heat the butter and oil and
sauté the onion and garlic just until soft.
Add the remaining ingredients and whisk
over high heat until the mixture comes to a
boil.
    Reduce the heat and cook gently for 30
minutes, stirring frequently.

X-tra Sauces

Y's and Z's

Yams and sweet potatoes are fleshy tubers. Both are cooked the same way, but are of different families. The true yam is grown only in the tropics and is sweeter and moister than sweet potatoes. Sweet potatoes are better for baking. Choose firm, well-shaped and blemish-free yams or sweet potatoes of medium size.

Crunchy Sweet Potato Casserole

6 to 8 servings

6-8 sweet potatoes
⅓ cup butter
½ cup honey
2 eggs, beaten
½ cup milk
⅓ cup chopped pecans
⅓ cup shredded coconut
2 tablespoons flour
2 tablespoons butter, melted

Peel, halve and boil the potatoes in water until soft. Drain well and mash with a potato masher or with an electric mixer. There should be 4 cups.

Beat in the ⅓ cup butter, 2 tablespoons of the honey, the eggs and milk. When well blended, spread the mixture into a lightly buttered 1½- to 2-quart casserole dish.

Stir together the pecans, coconut and flour. Add the rest of the honey and the melted butter to the nuts and blend well. Spread the mixture over the sweet potatoes.

Bake at 325 degrees F. for one hour.

Yam Flambeau

4 servings

4 medium yams
3 tablespoons butter
Sugar
Rum or bourbon

Boil the yams in water for 20 to 25 minutes or until tender. Drain.

When cool, peel and slice. Melt the butter in a heavy skillet and sauté the yam slices until brown on both sides.

Sprinkle with a little sugar and rum or bourbon and touch with a lighted match just before serving.

Deep Dish Yams

4 to 6 servings

5 tablespoons butter
1 medium onion
3 cups boiled yams, peeled and sliced
2 cups apple chunks
½ cup diced celery
¾ cup brown sugar
1 teaspoon lemon juice
1 teaspoon salt or 1 teaspoon
  dried basil
¼ teaspoon white pepper
2 tablespoons rum (optional)

Heat 2 tablespoons of the butter and sauté the onion until tender. Melt the remaining butter in a small saucepan.

Butter a deep 8-inch baking dish and put in a layer of the yams. Cover them with some apple chunks and celery and sprinkle with a little onion, half the brown sugar, lemon juice, salt and pepper and melted butter.

Repeat the process using the rest of the ingredients and finish with a layer of apples and brown sugar. Sprinkle with lemon juice and some rum, if desired.

Bake uncovered at 350 degrees F. for 30 minutes.

Yams,
Sweet Potatoes

Tennessee Sweet Potato Pudding

4 to 6 servings

3 eggs
3 cups grated sweet potatoes
½ cup sugar
½ teaspoon salt or 2 teaspoons
   lemon juice
½ cup raisins
½ cup currants
¾ cup chopped walnuts
¾ cup milk
½ cup molasses
¼ cup corn syrup
1 teaspoon cinnamon
½ teaspoon nutmeg
¼ teaspoon cloves
4 tablespoons butter
Heavy cream

Beat the eggs until light and mix in the sweet potatoes thoroughly. Add all the ingredients except the butter and cream and blend well.

Heat the butter in a heavy iron skillet or in an ovenproof serving dish. When bubbling stir in the sweet potato mixture. Continue stirring until the mixture begins to boil.

Bake at 350 degrees F. for 45 minutes, stirring every 15 minutes.

Serve with a pitcher of rich, heavy cream.

Zucchini, yellow crookneck and straightneck, scallop and chayote are types of summer squash. All are of the gourd family and grow on vines, with the exception of the zucchini, which is a bush squash. Select fresh, small to medium-size squash that is heavy for its size.

Zucchini Appetizers

48 servings

4 small zucchini
1 cup biscuit mix
½ cup finely chopped onion
½ cup grated Parmesan cheese
2 tablespoons chopped parsley
½ teaspoon salt (optional)
½ teaspoon dried marjoram
1 clove garlic, minced
½ cup corn oil
4 eggs, slightly beaten

Wash and trim the ends of the young zucchini. Thinly slice them.

Mix all of the ingredients in a large mixing bowl. Spread the mixture into a well-greased 13x9x2 baking pan and bake at 350 degrees F. for 25 minutes, or until brown and bubbly.

Cut into small squares and serve as hot hors d'oeuvres.

Baked Zucchini and Summer Squash

4 to 6 servings

1 large onion, diced
1 large clove garlic, pressed
4 tablespoons butter
4 small summer squash
4 small zucchini
Salt and pepper (optional)
1 pint sour cream
1 cup fine bread crumbs
1 tablespoon chopped parsley

Cook the onion and garlic in 2 tablespoons of the butter in a large skillet just until the onion is soft. Wash and trim the ends of the young squash and zucchini. Slice them into the skillet and stir well.

Let them cook for 10 minutes. Cool completely. Season with salt and pepper, if desired, and stir in the sour cream.

Transfer to a shallow baking dish and cover with the bread crumbs and chopped parsley. Dot with the remaining butter.

Bake at 350 degrees F. for 30 minutes.

Zucchini

Zucchini Bread

1 loaf

3 eggs
1 cup safflower oil
2 cups sugar
1 teaspoon vanilla
3 small zucchini
3 cups flour
3 teaspoons cinnamon
¼ teaspoon baking powder
1 teaspoon soda
½ cup chopped nuts

Beat the eggs well and add in the oil, sugar and vanilla. Blend together well.

Wash, peel and grate the zucchini and beat into the egg mixture.

Sift together the dry ingredients and add them to the batter in three parts, beating well after each addition. Fold in the nuts.

Bake at 325 degrees F. in a well-greased loaf pan for about one hour or until a knife inserted comes out clean.

Cool for 10 minutes and remove from the pan. Cool completely before slicing.

Zucchini Cheese Bake

4 servings

2 medium zucchini, sliced
2 tablespoons melted butter
½ cup shredded sharp Cheddar cheese
1½ teaspoons basil
Freshly ground black pepper, to taste

Cover the bottom of a shallow baking dish with the zucchini slices. Pour over the melted butter. Distribute the cheese evenly over the zucchini slices and sprinkle the basil and pepper over top.

Bake at 375 degrees F. for 20 minutes.

Zucchini Toss

4 servings

1 tablespoon each butter and olive oil
1 medium onion, coarsely chopped
2 medium zucchini, sliced
1 garlic clove, pressed
2 large tomatoes, peeled and chopped
2 tablespoons grated Parmesan cheese
Freshly ground black pepper, to taste

Melt the butter in a medium-size skillet
and add in the olive oil. Stir in the onion,
cooking until just soft, but not brown.

Add the zucchini and garlic and cook
over medium heat for about 10 minutes,
stirring frequently. Test the zucchini with a
fork to make sure it is tender.

Add the tomato, stir and simmer over
low heat for 5 minutes. Sprinkle the
cheese and pepper over all. Toss and
serve immediately.

Duchess Stuffed Squash

8 servings

4 small summer squash
2 medium potatoes
2 tablespoons butter
2 egg yolks
Salt and pepper (optional)
1 teaspoon water
3 tablespoons grated Cheddar or
**   Gruyère cheese**

Wash the squash well. Cut off the tips at
both ends and slice lengthwise. Scoop out
the centers and discard. Boil the halves for
10 minutes. Drain well.

Peel and quarter the potatoes. Boil them
for 18 to 20 minutes or until tender. Drain
and return to the heat, tossing until dry
and mealy. Mash the potatoes, adding the
butter and one egg yolk. Season to taste
with salt and pepper, if desired.

Place the halves in an ovenproof serving
dish cut side up. Spoon the mashed
potatoes into the squash centers. Beat the
remaining egg yolk slightly with one
teaspoon of water. Paint the potato filling
with the egg yolk mixture. Sprinkle with the
grated cheese.

Bake 20 to 30 minutes at 300 degrees F.
Brown under the broiler and serve.

Zucchini

Index

Index